SpringerBriefs in Electrical and Computer Engineering

More information about this series at http://www.springer.com/series/10059

Michael G. Harvey

Wireless Next Generation Networks

A Virtue-Based Trust Model

Michael G. Harvey
Whiting School of Engineering
Johns Hopkins University
Pittsburgh, PA
USA

ISSN 2191-8112   ISSN 2191-8120 (electronic)
ISBN 978-3-319-11902-1 ISBN 978-3-319-11903-8 (eBook)
DOI 10.1007/978-3-319-11903-8

Library of Congress Control Number: 2014951722

Springer Cham Heidelberg New York Dordrecht London

Printed on acid-free paper

Springer is part of Springer Science+Business Media (www.springer.com)

For Missy and Kaija

Preface

The mobile Internet involves the ongoing convergence of fixed and mobile networks with other types of wireless next generation networks within an open and dynamic environment, where the lack of a central mediator forces entities to interact through collaboration and negotiation. Current computational approaches to trust, based on the iterative exchange of personal knowledge such as digital credentials and access control policies, are not feasible for wireless next generation networks due to the limited power, bandwidth, and computational capabilities of mobile devices. The lack of pre-authentication knowledge makes it difficult to establish initial trust between strangers. Use cases from secure routing and secure key management are used to illustrate the limitations of current computational trust models, and to motivate the need for a new trust model that reflects human social interaction and does not depend on personal knowledge of user identities.

A virtue-based trust model is proposed as an efficient and flexible version of trust without identity based on the actions of an entity rather than on personal knowledge of the actor. The trust model is influenced by recent contributions from virtue epistemology and the intuition that the actions of an entity should be more efficient to evaluate than the belief state of one entity regarding the intentions or future actions of another entity. From a theoretical point of view, a virtue-based trust model allows us to relate trust and rationality in a non-circular fashion by showing how trust and reason are complementary cognitive mechanisms that guide our rational conduct at an animal or instinctual level and at a reflective level respectively. From a practical point of view, in addition to protecting confidentiality, a virtue-based trust model can help ensure availability if used to support local information sharing schemes. Furthermore, given its emphasis on virtue and character as universal traits of trustworthiness and its moderating notion of achieving balance or harmony in the trust relation, a virtue-based trust model can be adapted to the intercultural context of the mobile Internet.

The trust model integrates the behavioral and cognitive dimensions of trust in a nomological framework which includes both objective and subjective decision structures. These decision structures correspond to a distinction between animal knowledge and reflective knowledge respectively. Initial trust in other entities is established at the level of animal knowledge through innate competences called basic trust dispositions that are rational but not reason-based. Basic trust involves two pre-reflective and mutually dependent trust dispositions, self-trust and trust in other people. The trust relation involves the mutual adjustment, moderation, and self-control of one entity's basic trust dispositions or reactions in response to another entity's actions with the goal of achieving balance or harmony in the trust relation, whereby each entity accepts more or less the same degree of risk or vulnerability. For most interactions, the trust relation can be evaluated according to an objective decision structure using simple rules based on adapting actions and adjusting actions. Whereas adapting actions increase the level of trust and facilitate cooperation and collaboration through mutual adjustment, adjusting actions increase the level of distrust and lead to selfishness and conflict through the domination of one side by the other side. At the level of animal knowledge, the trust model reflects how humans do what they do naturally.

When there is a violation of initial trust, the trust relation needs to develop from basic trust toward full-fledged trust at the level of reflective knowledge. As the awareness of risk or vulnerability in the trust relation increases, an entity needs to justify its initial trust in another entity by establishing the reliability of the source. If the level of trust falls below a minimum acceptable threshold, where selfish behavior may lead to conflict, the trust relation needs to be evaluated according to a subjective decision structure using more complex rules based on acts of intellectual virtue that manifest the reliability or trustworthiness of an entity. At the level of reflective knowledge, the trust model reflects how humans can do a better job of what they do naturally by exercising reason-based competences such as intellectual virtues which help them avoid being either too trusting of other entities or not trusting enough of them.

A virtue-based trust model should be more efficient than identity-based, role-based, or attribute-based trust models that depend on computationally expensive methods. We need not evaluate the belief state of one entity regarding the intentions or future actions of another entity based on personal knowledge of the actor. Instead, we can evaluate the behavioral and cognitive performances of autonomous rational agents, whether human or artificial. An apt behavioral performance can be defined as an adapting action that enhances basic trust in other people and facilitates cooperation and collaboration. In contrast, an apt cognitive performance can be defined as an act of intellectual virtue that achieves balance or harmony in the trust relation by moderating the basic trust dispositions in the presence of the awareness of risk or vulnerability. Thus, we can evaluate the success of adapting actions in achieving balance or harmony in the trust relation so long as there is no violation of the trust relation. When a trust violation occurs, we can evaluate the success of acts of intellectual virtue manifested by one entity in moderating its own basic trust dispositions or reactions in response to the actions of another entity according to whether the acts achieve balance or harmony in the trust relation.

Whereas the present work aims to develop a theory of trust that reflects human social interaction, future work will need to define the methods of the theory and illustrate their application in different areas of network security such as secure

routing and secure key management. An ontology needs to be defined for orga-
nizing adapting actions and acts of intellectual virtue in a nomological framework.
Simple rules need to be formulated for distinguishing adapting actions from
adjusting actions in accordance with how successful they are in achieving the
socially valuable ends of cooperation and collaboration, while avoiding the socially
undesirable ends of selfishness and conflict. More complex rules need to be formulated for determining whether an entity has manifested a certain intellectual
virtue or reason-based competence that is conducive to increasing the level of trust
in a given interaction. Finally, potential defeaters of the methods need to be considered. In particular, acts of intellectual virtue may not be sufficient for distinguishing atypical interaction scenarios such as malicious hacking versus ethical
hacking, since both actions involve the exercise of similar intellectual virtues and
we have to take into account the intention of the action.

The interdisciplinary nature of a virtue-based trust model should appeal to at
least two different groups of researchers. As the problem of developing more
efficient and flexible trust models for wireless next generation networks has become
more pressing in computer science, the nature of trust and its role in society have
emerged as topics of widespread interest in philosophy and the social sciences.
Thus, on one hand, computer scientists may benefit from the normative, sociological, and cultural analyses of trust provided by philosophers and social scientists
in developing more efficient and flexible trust models that reflect the way humans
interact in social environments. Philosophers and social scientists, on the other
hand, may benefit from the empirical application of trust models in computer
science in understanding the practical limitations of their own theories and models
of trust. Consequently, a virtue-based trust model may best be seen as an example
of experimental philosophy that aims to make a small theoretical contribution to
computer science.

Pittsburgh, PA   Michael G. Harvey

Acknowledgments

The author would like to thank Dr. Harold J. Podell, Assistant Director of IT Security in the Center for Science, Technology and Engineering at the Government Accountability Office, for directing his attention to key management issues in wireless next generation networks. These issues, among others, have motivated the need for a more efficient and flexible trust model. The author would also like to thank Dr. Ernest Sosa, Board of Governors Professor of Philosophy at Rutgers University, for introducing him to epistemology as a doctoral student at Brown University. With respect to both influences, however, the author bears full responsibility for the proposal of a virtue-based trust model, and for his interpretation of Sosa's virtue perspectivism which provides the theoretical anchor for the trust model.

Most importantly, the author would like to thank his spouse, Melissa J. Harvey, for her research assistance in verifying the references, obtaining copyright permissions for material used in the book, and for encouraging him to think outside the box and perhaps even against the grain.

The author would like to thank the publishers for copyright permissions to reproduce material from the following papers:

©2009 Elsevier. Fig. 1.1 is reprinted with permission from TalebiFard P et al. (2010) Access and service convergence over the mobile internet–A survey. Computer Networks 54(4):545–557.

©2012 CRC Press. Fig. 2.2 is reprinted with permission from Chen L, Gong G (2012) Wireless security: Security for mobility. In: Communication system security. CRC Press, New York.

©2008 ACM Press. Figs. 2.3–2.5 are reprinted with permission from Hoeper K et al. (2008) Security challenges in seamless mobility–How to 'handover' the keys? In: 4th international ACM wireless internet conference (WICON '08), Maui, HI, 17–19 November 2008, pp 2–3.

©2013 IEEE. Figs. 2.6–2.7 are reprinted with permission from Kishiyama Y et al. (2013) Future steps of LTE-A: Evolution toward integration of local area and wide area systems. IEEE Wireless Communications 20(1):12–18.

©2013 NTT DOCOMO, Inc. Figs. 2.8–2.9 are reprinted with permission from Zugenmaier A, Aono H (2013) Security technology for SAE/LTE (system architecture evolution 2/LTE). NTT DOCOMO Technical Journal 11(3):28–30, 2013.

©2011 Emerald Group Publishing. Fig. 6.6 is reprinted with modifications and permission from Du R et al. (2011) Integrating Taoist yin-yang thinking with western nomology: A moderating model of trust in conflict management. Chinese Management Studies 5(1):55–67.

Contents

Figures

Tables

About the Author

Michael G. Harvey holds an undergraduate degree magna cum laude in Physics and Astronomy from the University of Pittsburgh. He holds master's degrees from Yale University, Brown University, and Carnegie Mellon University, where he studied a variety of disciplines, including epistemology, theory of religion, cognitive psychology, and computer science. He is currently studying cybersecurity at Johns Hopkins University. His most recent publication on privacy and security issues for mobile health platforms, co-authored with his spouse Melissa J. Harvey of the National Network of Libraries of Medicine, Middle Atlantic Region, appeared as the lead article in the July 2014 issue of *Journal of the Association for Information Science and Technology (JASIST)*.

Chapter 1
Introduction: Motivations for a New Trust Model

Trusted computing in mobile platforms has become a growing concern as mobile applications have evolved from voice communications to software applications provided by the mobile Internet. The evolution of the mobile Internet has exposed mobile devices to greater risk as security threats against wired networks have spread to wireless networks. In closed, fixed networks, computer-mediated interactions tend to be one-way. Fixed entities are granted privileges to access the assets of an organization within a closed network perimeter, and risk or vulnerability resides mostly with the organization. Traditional approaches to network security assume that the asset provider is trustworthy, that the size of the user population is small, that users can be authenticated based on personal knowledge of their identities, and that users gain trust once they are authenticated.

These assumptions, however, do not hold for open, dynamic environments such as the mobile Internet. As wireless communications have increased user populations, the task of maintaining the identities of all potential users is not feasible. Computer-mediated interactions, moreover, now involve the mutual exchange of personal knowledge such as digital credentials and access control policies, where both entities in the interaction are exposed to risk or vulnerability. The exchange is typically conducted by autonomous rational agents acting on behalf of unknown entities, which can be human or artificial, and often involves significant overhead in the computation of trust levels and expected outcomes in the trust relation. The problem of establishing initial trust in an open, dynamic environment, as well as the problem of moderating trust by enhancing or restraining it depending on the actions of agents, have become leading topics of research in network security.

To address these problems, researchers argue that next generation approaches to network security will need to model the way people interact in social environments. This will require an understanding of the sociological context and philosophical nature of trust. Misztal [1] argues that the renewed interest in the concept of trust can be attributed to the emergence of a widespread acknowledgement that existing bases for social cooperation, solidarity, and consensus have been eroded. Most current sociological approaches to this problem place a strong emphasis on moderation, mutual adjustment, and self-control based on trust. Although the concept of trust is attractive as a basis for social cooperation and integration, Misztal points out

© The Author(s) 2014
M.G. Harvey, *Wireless Next Generation Networks*,
SpringerBriefs in Electrical and Computer Engineering,
DOI 10.1007/978-3-319-11903-8_1

that it is difficult to examine due to its circularity. While it is argued that trust is one of the important sources of cooperation, it is also argued that trust is a socially valuable end. To achieve this end, one needs to proceed in a trustworthy way. Thus, what is needed is a non-circular theory of trust, one that by integrating the notions of trust and rationality in a unified conception reflects how people interact in social environments.

1.1 The Cybersecurity Context of the Mobile Internet

The mobile Internet involves the ongoing convergence of wireless next generation networks (NGNs) with internet protocol (IP)-based networks. The fixed/mobile convergence (FMC) will offer many practical benefits in enabling users to roam between fixed and mobile networks that employ different network access technologies. This flexibility, however, will come at the cost of enhanced techniques for performance and security. Wireless IP-based NGNs will not only create new technical problems for mobility management and quality of service (QoS) such as the ability to change access point or base station and QoS guarantees between different networks and network security domains. They will create new privacy and security challenges that require the coordination of different security policies and security levels across diverse networks.

Ross et al. [2] provide general security recommendations for both wired and wireless networks, and Souppaya and Scarfone [3] provide more specific security recommendations for mobile devices. Existing security threats to wired and wireless networks, as well as new privacy and security challenges posed by the mobile Internet, need to be addressed through a comprehensive cybersecurity framework [4]. Although the framework is intended for organizations responsible for protecting critical infrastructures vital to national security, economic security, and public health or safety, it can be applied to any organization that needs to protect sensitive assets. The framework aims to establish a consistent and iterative approach to managing cybersecurity risk through risk management, which involves the ongoing process of identifying, assessing, and responding to risk or the perception of vulnerability. Individuals and organizations need to understand the likelihood that a threat will occur and its potential impact on individual or organizational assets. In the light of this knowledge, individuals and organizations can determine an acceptable level of risk or vulnerability which can be expressed as their risk tolerance. This definition of risk as the level of vulnerability which an entity or organization is willing to accept is assumed in subsequent chapters.

To help identify the attack vectors or threats that have to be addressed by a cybersecurity strategy for networks connected to the mobile Internet, the key technical and security requirements that need to be met in the ongoing evolution of the mobile Internet can be summarized. TalebiFard et al. [5] argue that the mobile Internet will involve convergence at the network, service, and application levels. Network access convergence will involve the integration of different mobile networks with fixed networks in a single IP-based multi-service infrastructure. Service convergence will involve the management of services over such a common networking platform through a set of shared mechanisms. The key enabling technology for service convergence will be an all-IP core network enhanced by the IP multimedia subsystem (IMS). IMS is itself a converged network technology that enables users of third generation (3G) cellular networks to access packet-switched multimedia services over IP-based networks. IMS is designed to support IP-based multimedia communications independently of underlying network access technologies and characteristics of user terminals. IMS will facilitate access to the mobile Internet by supporting user and terminal mobility. As such, IMS will also serve as a common platform to support the development of multimedia applications and services for delivery to different terminal types over different network access technologies. Application convergence will involve the support of diverse applications through a common service platform enhanced by IMS. A potential service platform that can support application convergence is service-oriented architecture (SOA).

Figure 1.1 shows the three major functions performed by the service layer of the IP-based multi-service infrastructure. The service delivery platform (SDP) abstracts the complexity of the underlying network infrastructure from the application layer through a set of APIs. The centralized policy management function accepts or rejects requests for network applications, resources, and services based on business

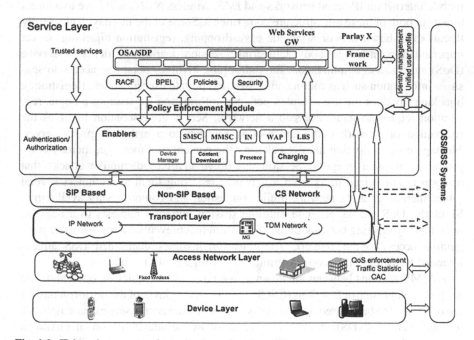

Fig. 1.1 IP-based core network service layer functions [5]

rules, user profiles, and network resource availability. Service enablers perform tasks that are common to most applications. The policy enforcement (PE) module provides centralized access control and management of network applications, resources, and services. When an unknown entity requests access to a network application, resource or service, the PE module performs authentication and authorization. In addition to tracking the addition, modification, or removal of network applications and resources and the policies associated with them, the PE module can accept, reject, terminate, or negotiate changes to service requests based on the relevant policy. This decision-making process requires information from other modules within the service layer, including user service subscriptions, the user service class, any service level agreements (SLAs) between the network service provider and third-party service providers, user account status, and user personal data and preferences. The PE module should retain a log of its policy decisions, and should record all events such as service requests, responses or actions, and errors. These data can be relayed to the operations support system (OSS) for troubleshooting and improved service. The PE module enforces three types of policy rules, including service policy rules, user policy rules and network policy rules, where each policy rule can be executed by a policy engine or service broker. Policies and rules can be communicated to other entities through the generic user profile (GUP).

The convergence of IP-based core networks with heterogeneous wireless access networks raises several privacy and security issues. Given the dependence of the mobile Internet on IP-based networks and IMS, wireless NGNs will have to address threats to both of these infrastructure components. Some of the more typical security threats to IP-based networks include eavesdropping, registration hijacking, server impersonation, message body tampering, session tear-down, denial-of-service (DoS) attacks, and amplification. Eavesdropping involves obtaining access to sensitive information such as routing addresses and private user identities. Registration hijacking involves the use of stolen personal information for sending a bogus registration request to gain access to a network. Server impersonation involves the redirection of network traffic and routing requests to a malicious proxy server. Message body tampering involves altering the content of a message, thereby compromising its data integrity. Session tear-down involves disruptive attacks that disconnect authorized users from a network, requiring them to reconnect several times, thereby creating network traffic congestion and compromising availability. Similarly, DoS attacks generate multiple session initiation requests to a network host, thereby creating network traffic congestion that prevents authorized users from gaining access to the network. Amplification involves distributed DoS attacks intended to increase the scope, magnitude, and impact of DoS attacks.

All IMS sessions between unknown as well as known entities on one hand, and the call session control functions (CSCFs) in the service layer of the infrastructure on the other need to be secured. IMS security issues include access security and network domain security (NDS). Access security involves the authentication of unknown entities and networks, as well as the authentication of the network traffic between them. In a hop-by-hop approach, where the address of the next device or hop along a given path in a network is listed in a routing table, access security can be considered the first hop between an unknown entity and the proxy-CSCF or gateway. User equipment (UE) needs to communicate with the serving-CSCF for authentication and authorization and the establishment of their IPsec security associations using register transactions. The serving-CSCF downloads authentication information for an entity from the home subscriber server (HSS). The proxy-CSCF subsequently establishes the IPsec security associations with the UE terminal. The UE also needs to authenticate the network to ensure that it is not counterfeit.

Network domain security involves protecting network traffic between the network nodes or CSCFs, which are globally protected, and different security domains which may belong to the same or different operators. All network traffic should traverse a security gateway (SEG). All entities exchange network traffic with SEGs within a security domain using IPsec, regardless of whether they reside outside or inside the security domain. Session initiation and termination are controlled by the broadcast multicast service center (BM-SC) which contains the security functionality for the IMS multimedia broadcast multicast service (MBMS). The BM-SC security function controls access to broadcast and multicast transmissions through data ciphering and key distribution. The UE needs to subscribe to a key request subfunction to obtain keys which are distributed by the MBMS key management function. The MBMS security framework, however, has not yet been generalized to address the privacy and security issues posed by the convergence of different network access technologies in fixed and mobile networks such as wireless local area networks (WLANs), 3G cellular, and WiMax. To minimize security silos at the interfaces of different network access technologies, fast and seamless key handovers across heterogeneous networks are required. This depends on a suitable key management scheme across these networks which is still unresolved for many security domains, including the integration of WLANs and wireless wide area networks (WWANs).

1.2 Limitations of Current Trust Models

A cybersecurity strategy needs to be developed to coordinate different security policies for different networks to reduce security silos at the interfaces of different network access technologies and security domains. In particular, the cybersecurity strategy needs to take into account that fixed networks and mobile platforms have different security levels and capabilities. Li and Hu [6] have recently conducted a study of wireless network security architectures that identifies three different protection strategies based on the notion of trusted terminals. Zheng et al. [7] have proposed a protection strategy that involves a trusted security architecture for 4G mobile networks based on trusted mobile platform (TMP) technology and public key infrastructure (PKI). Cheng et al. [8] have proposed a second protection strategy that involves a collaborative virus detection system for smart phones. The system performs joint analysis to detect single device and system-wide abnormal behavior based on the communication activity collected from the phones. Sailer et al. [9] have proposed a third protection strategy that involves a dedicated security server that can be added to the network to establish secure applications in terminals at the application service layer and secure access control policies at the network access layer. Curran [10] has proposed an additional protection strategy that requires network operators and software providers to deploy trusted intermediary technologies to prevent spam, spyware, and other intrusions. According to this protection strategy, mobile terminals can trust software applications that have passed security testing administered by a software testing service that ensures the software will not behave maliciously. Li and Hu [6] have elaborated trust relationships for such a software testing service for mobile users based on existing trust relationships in PKI.

Most of these protection strategies are based on the use of a centrally located supervisory control function to mediate trust between unknown entities. The power limitations of mobile platforms, however, preclude the use of a central mediator. In the open, dynamic networking environment of the mobile Internet, moreover, a universal trust model to control the interaction of entities through collaboration and negotiation is lacking. Instead, autonomous rational agents are used to interact, negotiate, and make agreements on behalf of unknown entities in a distributed environment. Automated trust negotiation (ATN) is a process that establishes trust gradually between two unknown entities by iteratively requesting and disclosing personal knowledge such as digital credentials and access control policies. Potential cooperative partners are evaluated and ranked according to their history of previous interactions with other entities. This allows agents to learn a probabilistic model of their behavior that can be used to predict the probability of future cooperation. A trust model can be developed from this probabilistic model to assist agents in selecting appropriate trust partners. Agents use trust models to compute the level of trustworthiness of a potential partner with whom they wish to cooperate, and perhaps eventually collaborate on activities of common interest.

In addition to the overhead involved in the computation of trust levels and likely outcomes of trust relations, the trustworthiness of entities can vary over time. This means that trust models need to be continuously updated with the most current knowledge about an entity. Ideally, this knowledge should consist of the most accurate and comprehensively coherent beliefs available about the entity's past behavior, credibility, and reputation. The gathering of such personal knowledge for all potential users of the mobile Internet is not feasible. These practical problems and limitations have motivated researchers to consider new ways of implementing a cybersecurity strategy to address the privacy and security issues posed by the mobile Internet. One approach that is receiving widespread attention is the use of real-time distributed control systems (RTDCSs) to monitor, control, and moderate the responses or actions of entities, based on knowledge about their behavior gathered in real-time by wireless sensor networks (WSNs). Such an approach to cybersecurity, however, requires a more efficient and flexible trust model that can be adapted in the intercultural context of the mobile Internet.

Since secure communication and access control in open, dynamic environments cannot utilize centralized authentication and authorization services, Galinović [11] argues that a decentralized, open trust model is needed, whereby entities can establish mutual trust on their own. Given the lack of a universal trust model for the mobile Internet, entities are forced to use different methods for authentication and authorization which create suspicion between entities. Furthermore, in most current computational trust models, all information needed for authentication and authorization is obtained during the pre-registration process rather than gradually as needed. Thus, if an entity wants to use n different Internet-based services, it has to undergo n separate registration procedures. In addition to having to remember n different authentication and authorization mechanisms, this situation increases risk or vulnerability since private and personal information is potentially stored in n different databases, thereby compromising confidentiality. Most importantly, people have shown that they value their privacy and are unwilling to identify themselves for every Internet-based service they want to use. Thus, Galinović argues that identity has not proven to be a useful basis for establishing trust between strangers. Instead, trust between strangers can be established on the basis of properties of a subject without requiring personal knowledge of the subject.

One approach to trust without identity that is gaining popularity is based on a socio-cognitive trust model which aims to integrate the cognitive and social aspects of trust. Villata et al. [12] have proposed such a trust model based on argumentation theory. According to this trust model, trust can be defined as the absence of attacks or arguments against a knowledge source or its information items on one hand, and as the presence of evidences or arguments in favor of the source or its information items on the other. The competence and sincerity of the source are two possible dimensions for evaluating its trustworthiness. Whereas competence concerns the extent to which a source is deemed able to deliver a correct argument, sincerity concerns the extent to which a source is deemed willing to provide such an argument. This socio-cognitive analysis of trust highlights two important properties of trust that any candidate trust model for wireless next generation networks should take into account. First, although initial trust in a source may be unjustified or have only minimal justification, rational trust ultimately needs to be defensible against attacks. To responsibly believe that a source of knowledge or its information items is trustworthy requires justified trust in the reliability of the sources and their deliverances. Secondly, the trustworthiness of a source depends on whether the knowledge or information items based on it has been generated in a competent or reliable fashion rather than carelessly and haphazardly.

Castelfranchi and Falcone [13] have developed these two subject-based properties of trust in another socio-cognitive theory of trust. In order to delegate a task to an agent, they argue that one has to believe that the agent is competent or able to do what is expected, and that the agent will actually achieve or help to achieve the desired end. Although one's trust in the agent can be represented by a belief state regarding the intentions or future actions of the agent, trust itself is not reducible to a belief state because it involves more than belief. In contrast to strictly cognitive and computational trust models, Castelfranchi and Falcone argue that trust is not reducible to the subjective probability, estimation, or belief that the agent will be able to do what is expected. Trust is a more dynamic process that also involves the awareness of vulnerability and its acceptance, which is accompanied by the feeling of being exposed to risk and uncertainty in one's belief about the trustworthiness of a source or its deliverances. Thus, trust involves a complex mental state or attitude of one agent toward another agent regarding a behavior or action that is deemed relevant for achieving or helping to achieve a goal they have in common. As such, trust can be seen as the mental or cognitive counterpart of delegation, which is an action performed by one agent who needs or desires the action of another agent in achieving his or her plan. Like Villata et al., Castelfranchi and Falcone argue that basic trust in other people needs to be articulated in beliefs and supported or justified by other beliefs to be fully rational. The credibility or trustworthiness of a belief depends on the credibility or trustworthiness of its source. According to a cognitive model of trust, the sources or bases of trust are beliefs consisting of evaluations and expectations. What matters is an agent's confidence or degree of justified trust in those beliefs.

Perhaps the most significant contribution of Castelfranchi and Falcone's socio-cognitive approach to trust is their conception of rational trust, or what it means to be properly trusting. Rational trust is represented as the mean between two extremes of irrational trust. This approach is consistent with most current sociological approaches to trust which place a strong emphasis on moderation, mutual adjustment, and self-control in the trust relation. When one agent trusts another agent for its source of knowledge, the trusting agent can have too much trust or confidence in the other agent and ascent to belief too easily. Or it can have too little trust or confidence in the other agent and withhold ascent unnecessarily. Thus, one or both entities in the trust relation can be over-confident or over-diffident. In the case of over-confidence or being too trusting of other agents, the trusting agent is willing to accept too much risk, vulnerability, or ignorance. This can lead to several socially undesirable ends, including reduced control actions, additional risks, careless and inaccurate actions, distraction, delay in the repair of broken trust relations, possible partial or total failure in achieving the goal of cooperation, and additional cost for the recovery from any one of these socially undesirable ends. Most importantly, an excess of trust, which is characterized by no or little reflection and the lack of awareness of risk or vulnerability in the trust relation, can lead one to delegate to an unreliable or incompetent agent. It can also result in the lack of control over the other agent, or to including other agents in one's plan who do not have the ability to plan themselves or contribute useful actions to a plan.

In the case of over-diffidence or being too unwilling to trust other agents, the trusting agent accepts no or too little risk, vulnerability, or ignorance. This can also lead to several socially undesirable ends, most importantly the inability to trust which is characterized by an excessive degree of reflection and awareness of risk or vulnerability in the trust relation. The inability to trust can lead in turn to the inability to delegate or rely on good potential trust partners, reduced cooperation where good opportunities for collaboration are missed, a need for too many evidences, arguments, or good reasons for cooperation, and the creation of superfluous controls that cost time and intellectual or computational resources and lead only to conflict. Finding the mean between these two extremes of irrational trust involves determining an appropriate ratio or balance between over-confidence and over-diffidence. According to Castelfranchi and Falcone, the level of trust in another agent is appropriate when the marginal utility of the additional evidence for trusting the agent is inferior to the cost of acquiring it. Similarly, the level of trust for delegating to another agent for assistance in achieving one's plan is appropriate when the risk or vulnerability that one is willing to accept in case of failure is inferior to the expected subjective utility in case of success, taking all available delegations or actions into account.

The distinction between over-confidence and over-diffidence in trust relations is also found in the legal literature. Whereas legal policy seeks to maximize or minimize interpersonal trust in most cases, Hill and O'Hara [14] argue that a better strategy is to optimize interpersonal trust. This optimization argument corresponds to the conception of being properly trusting in Castelfranchi and Falcone's socio-cognitive approach to trust. Individuals can be either too trusting of other people or not trusting enough of them. Whereas insufficient trust results in foregone mutually beneficial opportunities, paranoia, and unnecessary tensions and conflict, an excess of trust results in ineffective monitoring, fraud, reduced efficiency, and incompetence. These two extremes, moreover, can result in trust relations that lead to socially undesirable ends with legal consequences. For example, people often trust members of their own social groups, while distrusting members of other social groups in ways that not only limit mutually beneficial interactions within a society but may lead to intentional or effective discrimination. The challenge lies in finding the appropriate balance or mean between being too trusting of other people and being too suspicious of other people in a particular interaction. Furthermore, Hill and O'Hara argue that legal scholars assume that trust and distrust, or doubt, cannot co-exist. In most trust relations, however, they observe that parties trust each other on some matters, while distrusting each other on other matters. In practice, developing a trust relation involves acquiring an overall sense of the trustworthiness of a person or agent, as well as a specific sense of how trustworthy the person or agent is in particular contexts. Thus, given the co-existence of trust and distrust in practice, legal policymakers should not take an all-or-nothing position regarding the desirability of interpersonal trust. These practical considerations strongly suggest that we do not always trust in accordance with an idealized theory of trust where trust and doubt are mutually exclusive.

Many of these properties of trust have also been developed in reliabilist theories of knowledge. In particular, some virtue epistemologists have given careful attention to the distinction between being too trusting (over-confident) and being too suspicious (over-diffident) of our own self as well as of other people. In general, virtue epistemology stresses the role of intellectual virtues in reliably acquiring, sustaining, and revising beliefs about oneself, other people, and the world. Other virtue epistemologists have developed a conception of responsible belief based on sufficient rationality that is analogous to Castelfranchi and Falcone's definition of trust based on utility. An agent is rationally justified in trusting another agent or source of knowledge when its reasons for trusting the source are good enough given a context. This conception of responsible belief, as well as the distinction between being too trusting and being too suspicious, will be developed in subsequent chapters in a virtue-based trust model.

Like socio-cognitive models of trust, a virtue-based trust model aims to address the limitations of strictly cognitive and computational approaches to trust in which beliefs are the measure of all things and identity systems are the means of measurement. In particular, a virtue-based trust model challenges the assumption in cognitive and computational models of trust of whether it is necessary in every case to evaluate belief states regarding the intentions or future actions of other agents, and how likely it is that these intentions will actually lead to the expected outcome of adapting actions that facilitate cooperation and collaboration. Most current theories of trust, including socio-cognitive models of trust, are based on logics of belief rather than on logics of action. Since modal logics enhance propositional or first-order predicate logics with modal operators that allow us to articulate beliefs, Rangan [15] argues that they have become the de facto standard for expressing trust relations and access control policies. But this strictly cognitive or axiomatic approach to trust involves a complex process that is computationally expensive and difficult to implement across heterogeneous networks.

Although beliefs play an important role in a virtue-based trust model, they are not the sources or bases for determining whether it is appropriate to trust other agents. Instead, trust in other agents is based on the actions performed by those agents in a particular interaction or social exchange. These actions are evaluable in terms of adapting actions that facilitate cooperation and collaboration and adjusting actions that lead to selfishness and conflict. This simplifies the approach to trust in most cases, while reserving recourse to more sophisticated negotiation and mediation strategies when the level of trust falls below a minimum acceptable threshold as determined by the context of the social exchange. But even in this case, as we will see in subsequent chapters, we need not evaluate belief states regarding the intentions or future actions of other agents based on personal knowledge of the agents. Instead, we can evaluate acts of intellectual virtue manifested by one entity in moderating its own basic trust dispositions or reactions in response to the actions of another entity according to whether the acts achieve in socially valuable ends such as cooperation and collaboration and avoid socially undesirable ends such as selfishness and conflict. Thus, a virtue-based trust model is a version of trust without identity that protects confidentiality.

References

1. Misztal BA (1996) Trust in modern societies: the search for the bases of social order. Blackwell, Cambridge, pp 1–8
2. Ross R et al (2013) Security and privacy controls for federal information systems and organizations. Joint Task Force Transformation Initiative Interagency Working Group, NIST, Special Publication 800-53, rev 4. doi:10.6028/NIST.SP.800-53r4

3. Souppaya M, Scarfone K (2013) Guidelines for managing the security of mobile devices in the enterprise. NIST, Special Publication 800-124, rev 1. http://www.nist.gov/customcf/get_pdf.cfm?pub_id=913427. Accessed 15 Jul 2014

4. NIST (2014) Framework for improving critical infrastructure cybersecurity. NIST, ver 1, pp 3–5. http://www.nist.gov/cyberframework/upload/cybersecurity-framework-021214.pdf. Accessed 15 Jul 2014

5. TalebiFard P et al (2010) Access and service convergence over the mobile internet–a survey. Comput Netw 54(4):545–557. doi:10.1016/j.comnet.2009.08.017

6. Li T, Hu A (2013) Trust relationships in secured mobile systems. In: 2013 IEEE wireless communications and networking conference (WCNC 2013), Shanghai, China, 7–10 Apr 2013, pp 1882–1887

7. Zheng Y et al (2005) Trusted computing-based security architecture for 4G mobile networks. In: 6th international conference on parallel and distributed computing, applications and technologies (PDCAT 2005), Dalian, China, 5–8 Dec 2005, pp 251–255

8. Cheng J et al (2007) SmartSiren: virus detection and alert for smartphones. In: 5th international conference on mobile systems, applications and services (MobiSys '07), San Juan, PR, 11–14 Jun 2007, pp 258–271

9. Sailer R et al (2004) Design and implementation of a TCG-based integrity measurement architecture. In: 13th conference on USENIX security symposium, San Diego, CA, 9–13 Aug 2004, pp 223–238

10. Curran CD (2006) Combatting spam, spyware, and other desktop intrusions: legal considerations in operating trusted intermediary technologies. IEEE Secur Priv 4(3):45–51. doi:10.1109/MSP.2006.60

11. Galinović A (2010) Automated trust negotiation models. In: 33rd international convention on information and communication technology, electronics and microelectronics (MIPRO 2010), Opatija, Croatia, 24–28 May 2010, pp 1197–1202

12. Villata S et al (2013) A socio-cognitive model of trust using argumentation theory. Int J Approximate Reasoning 54(4):551–556. doi:10.1016/j.ijar.2012.09.001

13. Castelfranchi C, Falcone R (2005) Socio-cognitive theory of trust. In: Pitt J (ed) Open agent societies: normative specifications in multi-agent systems. Wiley, New York, pp 58–89

14. Hill CA, O'Hara EA (2006) A cognitive theory of trust. Wash Univ Law Rev 84(7):1717–1796. doi:10.2139/ssrn.869423

15. Rangan PV (1992) An axiomatic theory of trust in secure communication protocols. Comput Secur 11(2):163–172. doi:10.1016/0167-4048(92)90043-Q

Chapter 2
Wireless Threats and Key Management Issues

We have discussed how the mobile Internet involves the ongoing convergence of wireless next generation networks (NGNs) with IP-based core networks, and some of the more typical security threats to both components of the infrastructure. The evolution of the mobile Internet also involves the ongoing convergence of different wireless technologies such as third generation (3G) mobile networks, wireless local area networks (WLANs), wireless wide area networks (WWANs), mobile WiMAX, and wireless sensor networks (WSNs). Although WSNs were motivated by military applications for surveillance and national defense, they are increasingly being deployed in a variety of civilian applications such as medical devices, home healthcare, autonomous vehicles, smart structures, supervisory control and data acquisition (SCADA) systems, disaster management, and cybersecurity using real-time distributed control systems (RTDCSs). Similarly, although RTDCSs were motivated by the need to monitor national critical infrastructures such as electric power grids, oil and natural gas supplies, water and wastewater distribution, and transportation systems, they are increasingly being deployed in cybersecurity to protect portions of critical information infrastructures such as wireless networks due to their ability to monitor the activity in a localized area or region.

Many of the privacy and security issues of WSNs and RTDCSs also hold for wireless NGNs in general. Despite their increasing popularity, wireless NGNs remain susceptible to several security threats due to their distributed nature and to the limited computational capabilities and energy supply of mobile nodes and sensor nodes. Typical attack vectors in all types of wireless technologies include threats that are common to both wired and wireless networks, jamming attacks against wireless communication channels, and attacks aimed at key distribution in mobile networks with roaming agreements that have different network access technologies such as WLANs and WWANs. Any trust model supporting the convergence of fixed and mobile networks needs to address these attack vectors by being able to respond to different security policies for each type of network and to different security domains in a flexible fashion. Such a trust model also needs to address the problem of key management for mobility in wireless NGNs to control entities as they roam within the same network or between different types of networks. In particular, methods for secure key derivation and distribution should

© The Author(s) 2014
M.G. Harvey, *Wireless Next Generation Networks*,
SpringerBriefs in Electrical and Computer Engineering,
DOI 10.1007/978-3-319-11903-8_2

avoid approaches that have large computation and communication overhead to
expedite fast and seamless handovers to minimize data leakage and prevent network
service disruptions for authorized users.

2.1 Attack Vectors in Wireless NGNs

In general, whereas traditional wired networks can have distant attackers but adequate perimeter defenses, wireless networks need to be within range of an attacker's
wireless device and can be accessed without a perimeter defense. Kisner et al. [1]
argue that the security requirements for WSNs can be divided into primary goals
and secondary goals. Primary goals include confidentiality, integrity, authentication, and availability. Secondary goals include data freshness, self-organization or
distributed collaboration between nodes, node synchronization, and secure locali-
zation of secret information such as keys. Similarly, typical attacks against wired
and wireless networks can be divided into passive attacks carried out by selfish
nodes and active attacks carried out by malicious nodes. Passive attacks include
monitoring of network traffic and wireless communication channels for information
that can be used to execute active attacks. These attacks typically involve eavesdropping and traffic analysis. In eavesdropping, an attacker acquires data passively
by intercepting data exchanges. This type of attack includes decrypting any
encrypted data. In traffic analysis, an attacker deduces properties of a data exchange
based on personal knowledge of the interacting entities, the duration of the data
exchange, timing, bandwidth, and other technical characteristics that are difficult to
disguise in packet networks.

Active attacks include masquerading, replay attacks, message modification,
denial-of-service (DoS), jamming, and routing attacks. In masquerading attacks, an
attacker impersonates an authorized entity to gain access to network applications,
resources, or services. Man-in-the-middle attacks involve a double masquerade,
where the attacker convinces the sender that she is the authorized recipient of a
message on one hand, and convinces the recipient that she is the authorized sender
of the message on the other. In replay attacks, an attacker injects malicious packets
into the network to disrupt it. Typically, an attacker rebroadcasts a previous message to cause the network to reset, thereby placing it in a vulnerable state, and to
gather information for further attacks. Replay attacks most often compromise
integrity, but can also compromise authentication, access control or authorization,
and non-repudiation. In addition, selective replay attacks can negatively impact
both availability and confidentiality. In message modification, an attacker alters
packets by inserting changes into them, deleting information from them, reordering
them, or delaying them. In wireless networks, message modification is typically
accomplished through man-in-the-middle attacks. This type of attack compromises
integrity, but can potentially affect all aspects of security. In DoS attacks, the
availability of network applications, resources, or services is compromised. In
wireless networks, this type of attack is typically accomplished by disabling one of

the interacting entities in the data exchange or by jamming the wireless communication channel.

Jamming refers to the disruption of a communications system such as a wireless network through the intentional use of electromagnetic interference. Jamming blocks a signal or message between two interacting entities by keeping the communications medium busy. An attacker sends a signal with a significantly greater signal strength relative to normal signal levels in the system to flood the channel. Thus, jamming is effectively a form of DoS attack. Jamming can be performed by a single attacker or multiple attackers working together, and can target a specific sender or receiver as well as the entire shared medium. A more sophisticated form of jamming involves the violation of the network protocol, where many more packets than normal are transmitted to increase the number of packet collisions. Wireless networks are more susceptible to jamming than wired networks because of their potential access from covert locations, and because these locations can be easily changed through mobility.

Routing attacks involve interference with the correct routing of packets through a network. Several different types of routing attacks can be carried out at the network layer, including spoofed, altered, or replayed routing information. These attacks can create routing loops, extend or shorten intended routing paths, generate bogus error messages, and increase end-to-end latency, thereby compromising availability. Selective forwarding attacks subvert a node in a network in order to drop selected packets. Similarly, sinkhole attacks subvert a node in a network in order to attract packets to it. Wormhole attacks record packets from one location in a network and retransmit them in another location in the network to disrupt its overall functionality.

Table 2.1 shows how Kisner et al. have summarized these attack vectors in a cybersecurity attack-vulnerability-damage (AVD) model. The AVD model expresses attack vectors as malicious actions that target a vulnerability in a network whose impact affects some aspect of its performance. The model also shows how these attack vectors can be organized according to the level of their severity. Wireless network designers can use the AVD model to assist them in developing a defense-in-depth or layered approach to security.

Current protection strategies against these attack vectors are inadequate. In the case of WSNs, where sensor nodes are often deployed in adversarial or hostile environments, the sensors themselves need to be physically protected to prevent them from being replaced with malicious nodes. In the case of the wireless sensor network itself, several different trust models for secure routing have been proposed to minimize the impact of these attack vectors. Vasilache et al. [2] have compared two popular trust models for WSNs, μRacer and k-FTM. The μRacer trust model, proposed by Rezgui and Eltoweissy, is an adaptive and efficient routing protocol suite designed for sensor-actuator networks as shown in Fig. 2.1. In a sensor-actuator network, distributed controllers, sensors, and actuators are connected in a wireless sensor network to monitor and control the correct operation of a physical system. Information about the environment of the physical system is collected by sensors and serve as inputs to the controllers. Controllers perform corrective actions

Table 2.1 Cybersecurity attack-vulnerability-damage model [1]

Attack			Vulnerability	Damage		
Origin	Action	Target		State effect	Performance effect	Severity
Local	Probe	Network	Configuration	None	None	None
Remote	Scan	Process	Specification	Availability	Timeliness	Low
	Flood	System	Implementation	Integrity	Precision	Medium
	Authenticate	Data		Confidentiality	Accuracy	High
	Bypass	User				
	Spoof					
	Eavesdrop					
	Misdirect					
	Read/Copy					
	Terminate					
	Execute					
	Modify					
	Delete					

Fig. 2.1 General architecture of a sensor-actuator network [1]

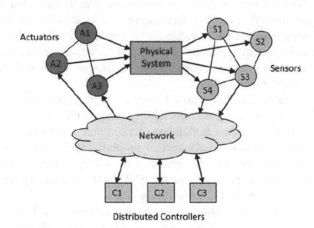

on the physical system through actuators connected to the system based on the values of sensor inputs.

The μRacer protocol suite consists of a trust-aware routing protocol (TARP), context-aware routing protocol (CARP), and a service-aware routing protocol (SARP). The TARP routing protocol aims to prevent packets from being routed through malicious or malfunctioning nodes using two concurrent sub-functions, reputation assessment and path reliability evaluation respectively. The reputation assessment sub-function assesses the direct reputation of a node based on its most recent interaction with another node, and the indirect reputation of the node based on its routing behavior or past interactions with its neighbor nodes. When an authorized node broadcasts a reputation request message regarding an unknown node, each node concerned broadcasts a reputation report on the unknown node. Upon receiving the reputation report, the requesting node updates the indirect reputation of the unknown node. If the reputation of an unknown node falls below a minimum acceptable threshold, any node in the network that is able to detect this situation broadcasts an unsolicited reputation report. The aggregate reputation of the unknown node is then calculated based on both its direct reputation and indirect reputation.

The k-FTM trust model, proposed by Srnivasan and Wu, is a k-parent flooding tree for secure and reliable broadcasting in wireless sensor networks. Based on the flooding tree model (FTM) for communications, where messages are sent from the root of the tree toward its leaves, this trust model aims to prevent denial-of-broadcast message (DoBM) attacks which are similar to DoS attacks. The wireless network topology of this trust model is well suited to broadcast communications such as the IP multimedia subsystem (IMS) which will serve as the key enabling technology for service convergence in the all-IP core network of the mobile Internet. Except for nodes that are within transmission range of the base station, each node in the tree has exactly k parents. Blind flooding, where each node rebroadcasts a message when it is received for the first time, is carried out once to generate the initial structure of the flooding tree. Each node in the flooding tree retains a reputation value for each of its neighbors, namely, its parent and child nodes. Messages are broadcast in two phases which include the broadcast phase and the acknowledgment phase. Both phases are susceptible to routing attacks, however, which can employ a compromised node to block a message, thereby preventing it from reaching or flooding the entire sub-tree. Thus, both tree height and node degree are important in preventing DoBM attacks.

As usual in any engineering field, these trust models involve tradeoffs. When computing the trust value for an unknown sensor node, μRacer takes into account the indirect reputation of the node as reported by its neighbor nodes in addition to its direct reputation. In contrast, k-FTM imposes an extra topology characteristic to ensure that messages can flow in the tree in case of a malicious or malfunctioning node. Whereas trust values computed by μRacer have the least fluctuation over time, thereby enhancing security, k-FTM is able to determine more quickly that a node is compromised by malicious or malfunctioning behavior, thereby enhancing availability. Thus, while k-FTM is faster at detecting compromised nodes, its trust values for unknown nodes are more uncertain given greater fluctuations due to taking into account only the direct reputation of the node. Conversely, while μRacer is slower at detecting compromised nodes due to the additional computation time required to calculate the indirect reputation of an unknown node, its trust values for the node are more certain. This comparison suggests that μRacer is more suitable for WSNs where routing attacks are of primary concern, whereas k-FTM is more suitable for WSNs that are deployed in harsh environmental conditions where nodes may fail. Both trust models, however, illustrate limitations that motivate the need for a new trust model for wireless next generation networks. Whereas μRacer has a large trust value computation overhead, k-FTM has a large communication overhead. To achieve a better balance between security and availability, we need a trust model that reduces both the computation overhead in the calculation of trust values for nodes and the communication overhead in the number of messages that need to be exchanged between nodes.

2.2 Key Management for Mobility in Wireless NGNs

Barker et al. [3] argue that cryptography is used to protect information from unauthorized disclosure, to detect unauthorized message modification, and to authenticate the identities of network entities. A network entity can be a human agent, including an individual or an organization, or it can be an artificial agent such as a network device, process, or autonomous rational agent acting on behalf of a human agent. A cryptographic key management system (CKMS) protects keys and metadata from being stolen and decrypted. In addition to managing and protecting cryptographic keys throughout their lifecycles, a CKMS needs to protect certain metadata about the key such as the cryptographic algorithm it uses, the authorized uses of the key, and the security services that are provided by the key.

Since keys can be stolen by an attacker when they are generated by a serving network and distributed to a target network, key management in wireless next generation networks has become an attack vector in its own right. Barker et al. [4] argue that the proper management of cryptographic keys is essential for the effective use of cryptography. Poor key management can easily compromise the strongest cryptographic algorithms. Secure information protected by cryptography depends on the strength of the keys, the effectiveness of mechanisms and protocols that support the keys, and the protection of the keys themselves. All keys need to be protected from modification, and secret and private keys need to be protected from unauthorized disclosure or leakage. Effective key management provides the foundation for secure key generation, distribution, and storage. In the case of secure key generation and distribution, Bergstra and Burgess [5] argue that there is often a high level of uncertainty in knowing the true source of a cryptographic key. Beyond a certain threshold of evidence, one needs to trust the assumption of ownership. The higher the level of uncertainty regarding the source, the more risk or vulnerability one entity in a trust relation needs to accept. This can destabilize the trust relation between two interacting entities, based on the assumption that the risk or vulnerability accepted by both entities should be ideally shared or symmetric.

This problem has motivated the need for new approaches concerning how to securely handle key exchanges in mobility applications. When an entity switches from one point of attachment (PoA) to another in the same network or in a different network with a roaming agreement, the wireless network connection needs to be secure to prevent connecting to a rogue PoA or transmitting confidential data over an unprotected link. Hoeper et al. [6] describe the problem and propose three possible solutions. They begin by describing the requirements for full network authentication, which involve the establishment of a secure connection between a mobile node and PoA in a wireless network through an authentication and key exchange protocol. The authentication process involves the sharing or exchange of secret keys or passwords between two interacting entities during the protocol execution. At each security level or trust level in the wireless network, new keys are derived from the exchanged keys to protect subsequent communications between the mobile node and PoA.

The term "handover" originates from cellular networks, and refers to the process of changing the current PoA to a target PoA through the use of switches. According to Chen and Gong [7], a communications system is dynamic in that some nodes may move from one location to another as in the ubiquitous case of cellular phones. Such nodes are called mobile nodes which are terminals that can be connected to networks through different fixed network nodes or PoA as shown in Fig. 2.2.

Initially, the mobile node m is connected to the network through fixed node 2. As m moves closer to fixed node 5 in the network, the mobile node switches its connection from fixed node 2 to fixed node 5. The process of switching from one fixed node to another in the network is called a handover. If executed efficiently, handovers provide a continuous connection to the network through different base stations without disrupting communications. Service mobility refers to the protocols and mechanisms responsible for facilitating collaborative communication between mobile nodes. In order to handover key and service information from one base station to another, both entities in the data exchange must share the same knowledge about a cellular subscriber's roaming. The term "roaming" originates from the global system for mobile (GSM) communications standard used by mobile phones. Roaming agreements between networks help ensure that a traveling wireless device is kept connected to a network without breaking the connection. If a cellular subscriber travels beyond the transmitter range of his or her cell phone company, the cell phone should automatically hop onto another phone company's service, if available, using the subscriber's identity in the visited network. A handover is said

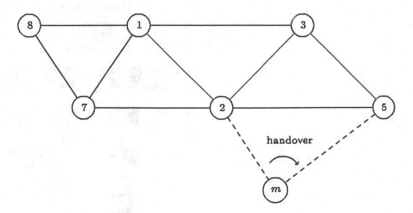

Fig. 2.2 A network with mobile nodes [7]

to be "seamless" when a new connection is established before the old one is broken or torn-down.

Hoeper et al. argue that the problem is how to execute the handover as quickly as possible to minimize security threats both to the network and to the mobile terminal, and to prevent network services from being disrupted while an entity is roaming. A common approach, called pre-authentication, establishes a full network authentication with a target PoA based on the current network connection before the handover is executed. A more efficient approach, called re-authentication, utilizes the information obtained from the establishment of keys in a previous authentication in the same network or in a different network with a roaming agreement to expedite the handover. The solutions proposed by Hoeper et al. are based on a re-authentication approach. Two different handover scenarios are considered, one involving a handover within a single security domain and the other involving a handover between two different security domains with a roaming agreement.

Figure 2.3 shows the architecture for a handover within a single security domain. A single security domain consists of one authentication server (AS) and several key holders connected to the authentication server. A mobile node (MN), which is currently attached to the serving PoA, plans to attach to a target PoA in the same security domain. Whereas the serving PoA is connected to the authentication server through $n - 1$ key holders, the target PoA is connected to the authentication server through a different path of $m - 1$ key holders. $L(x)[S]$-key holders and $L(x)[T]$-key holders refer to network entities in the serving and target networks respectively, which can serve as key holders in the authentication process. The key holder level in the network is represented by x. Lc-key holder is defined as the lowest common

Fig. 2.3 A handover within a single security domain [6]

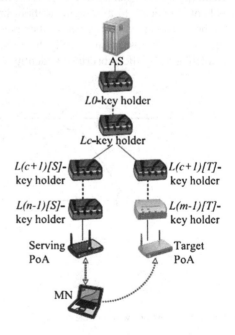

key holder before the paths from the authentication server to the serving and target PoAs split into two different branches.

Figure 2.4 shows the architecture for a handover between two different security domains that have a roaming agreement. Each security domain has its own authentication server, AS1 and AS2 respectively. In this case, the serving and target networks and their corresponding key holder paths are different. Typically, a key holder in the serving network can communicate with a key holder in the target network only through the authentication servers. A shortcut is defined as a direct communication path between a key holder in the serving network and a key holder in the target network that does not have to go through the authentication servers. Figure 2.4 illustrates a shortcut between the lowest common key holders, $Lc[S]$-key holder and $Lc[T]$-key holder.

In both handover scenarios, the handover assumes that there is a key hierarchy that has been established between a mobile node and the serving PoA in a previous authentication. Figure 2.5 shows the key hierarchy for the serving network S. The key hierarchy depends on the key holder path in the serving network S or the target network T, the wireless access technology i, and the number of key holder levels in the serving or target networks, denoted by n and m respectively. If the network access is successful, the mobile node and the authentication server derive the root key $RK_i[S]$. The authentication server then derives the $L0K_i[S]$ key for level 0 and sends it to the $L0$-key holder. The $L0$-key holder then derives the $L1K_i[S]$ key for level 1 from the $L0K_i[S]$ key and sends it to the $L1$-key holder. The key derivation and distribution process is continued until the lowest key holder in the chain, the PoA in the network, receives the $LnK_i[S]$ key. The PoA then derives the transient protection key $TPK_i[S]$ which is used by the mobile node

Fig. 2.4 A handover between two different security domains [6]

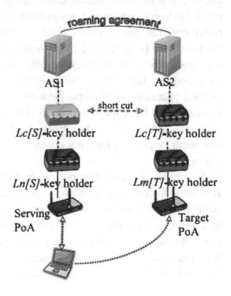

Fig. 2.5 A key hierarchy for
a wireless technology i [6]

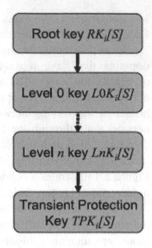

to derive all the keys in the hierarchy necessary to protect the wireless link
between itself and the PoA.

The security and availability issues associated with seamless handovers in
wireless next generation networks arise from two unresolved problems. First, unlike
cellular networks, other types of wireless networks lack a dedicated key management infrastructure to support the derivation of handover key hierarchies from
previous network connections and their distribution to a target network. Consequently, no dedicated network entities are available to trigger and manage key
distribution. Roaming information is necessary to trigger key distribution to the
correct target network. But in some wireless networks entities are unable to
exchange information about the roaming behavior of mobile nodes. Thus, it is
unclear how key distribution can be triggered. Once it is triggered, a network entity
needs to derive handover keys and distribute them to a target network. Hoeper et al.
point out that this network entity could be the serving authentication server, a
common key holder, or an entity with shortcut access to the target network. There
are other related problems that any suitable handover scheme needs to address,
including the fact that not all wireless networks share the same trust model. Two
entities operating in different networks or security domains at the same key holder
level may not have the same physical protection. Moreover, if the serving and target
networks have a different number of key holder levels, a given key holder level in
one security domain may not correspond to the key holder level in another security
domain.

Secondly, the key distribution protocol needs to be able to perform handover
functions in a timely fashion to prevent security threats and avoid network dis-
connections. As we saw in the comparison of two secure routing protocols, μRacer
and k-FTM, this means that these handover functions should not have a large
computation and communication overhead. At one extreme, if the network entity
that performs the handover functions is the serving authentication server, the
handover will be more secure but slower. At the other extreme, if the network entity

that performs the handover functions is an entity with a shortcut access to the target network, the handover will be faster but less secure. Many security features can be provided only through an authentication server such as network-wide key synchronization, homogeneous trust models, channel binding, and the prevention of replay attacks using sequence numbers or timestamps. Such security features increase the computation and communication overhead of the key distribution protocol which results in delays that could adversely affect availability.

Thus, as in the case of our analysis of secure routing protocols, there is a tradeoff between security and availability in handovers, depending on the key holder level at which the handover is executed. In the light of this tradeoff, Hoeper et al. suggest the following strategy. In the best case, where some security requirements can be limited or even suspended, a fast handover can be performed by using the lowest common key holder or a shortcut. Once the handover is complete, a full network authentication can be forced by limiting the key lifetime. This releases the serving network from liability and assures the target network that the new connection is secure. In the worst case, if no security requirements can be suspended or limited for any duration, both the serving and target authentication servers in the backbone need to be accessed during the handover. In this case, a timely initiation of the re-authentication protocol is required to minimize performance degradation.

2.3 Current Approaches to Seamless Handovers

Although mobile communication systems were originally developed for military applications, Forsberg et al. [8] discuss how the concept of a cellular network was extended to commercial applications in the early 1980s. The advanced mobile phone system (AMPS) and the Nordic mobile telephone system (NMT) were the first cellular networks developed in the U.S. and northern Europe respectively. These first generation systems utilized analog transmission techniques and frequency division multiple access (FDMA). These systems supported handovers between different cells in a network such as a phone call from a car. Second generation (2G) mobile systems appeared in the early 1990s, and were predominantly based on the global system for mobile (GSM) communications standard. These second generation systems utilized digital transmission techniques over the radio interface between the mobile phone and the base station and time division multiple access (TDMA). 2G systems provided increased network capacity due to the efficient use of radio resources, and supported improved audio quality and new types of security features due to digital coding techniques. Third generation (3G) mobile systems appeared in the early 2000s, and introduced the concept of fully mobile roaming. This concept makes it possible for users to access mobile services from anywhere in the world through a collaboration of standards bodies from Europe, Asia, and North America called the 3G partnership project (3GPP). 3G systems provided large increases in data rates up to 2 Mbps using wideband code division multiple access (WCDMA). GSM systems and 3G systems are divided into two different domains, the circuit switched (CS) domain for carrying voice and short messages and the packet switched (PS) domain for carrying data.

Long-term evolution (LTE) of radio technologies, together with system architecture evolution (SAE), were initiated by 3GPP a decade later as the next evolutionary step in mobile communication systems. The new system is called evolved packet system (EPS), and its most important component is the radio network called evolved universal terrestrial radio access network (E-UTRAN). The EPS system contains only a PS domain and provides large increases in data rates up to more than 100 Mbps by using orthogonal frequency division multiple access (OFDMA) for downlink traffic from the network to the mobile terminal and single carrier frequency division multiple access (SC-FDMA) for uplink traffic from the mobile terminal to the network. LTE is based on a radio interface specification standardized by 3GPP. The ongoing LTE standards development is progressing toward an enhanced LTE radio interface called LTE-advanced (LTE-A). The new radio interface is motivated by the need for higher communications system capability in the light of the growth of mobile data traffic due to the proliferation of smartphones and new mobile devices. Since the spectrum in lower frequency bands of the original LTE radio interface is becoming scarce, the new LTE-A radio interface requires the efficient utilization of higher frequency bands to sustain future growth and support further network densification.

Kishiyama et al. [9] argue that this requirement will involve the integration of wide and local area enhancements through multicell cooperation between macrocells and small cells. Figure 2.6 shows the future development of LTE. In LTE-A, standardized technologies for multicell cooperation include coordinated multipoint (CoMP) transmission/reception and enhanced intercell interference coordination (eICIC). To enable the upgrade from the original LTE radio interface to the new LTE-A radio interface, enhanced local area (eLA) specifications and technologies for the mobile

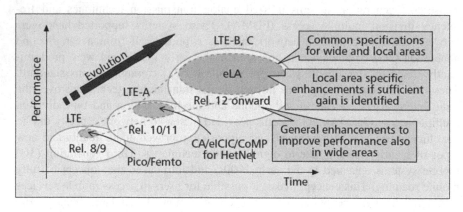

Fig. 2.6 Future development of LTE [9]

communications system need to be developed and added, while retaining common specifications for the LTE radio interface between wide and local areas.

Techniques are still needed to reduce the potential impact of the increased volume of signaling traffic on both the mobile network side and the mobile handheld device side. As network density is increased through the deployment of more small cells, these techniques need to optimize spectrum utilization which ranges from lower and narrower frequency bands to higher and wider frequency bands. Higher frequency bands, however, cannot be optimally utilized in wide areas of deployed macrocells due to space limitations on the evolved node B (eNB) side, which serves as a base station for the LTE radio access communications system. In addition, not only are higher frequency bands subject to higher path loss, but the cost of altering the existing network infrastructure to support higher frequency bands is prohibitive. Thus, the approach recommended by most researchers is to use lower frequency bands to provide basic coverage and mobility for macrocells in wide areas, and to use separate higher frequency bands to support small cells and high-speed data transmission in local areas. In contrast to wide areas consisting of macrocells, local areas refer to outdoor dense deployments and hotspots of small cells. Figure 2.7 shows how wide areas can be supported by lower and narrower frequency bands, while local areas can be supported by separate higher and wider frequency bands.

Kishiyama et al. argue that the integration of wide and local areas can be seen as a new form of cooperation between conventional macrocells in lower frequency bands and small cells in higher frequency bands. To support the integration, the authors introduce the novel concept of a phantom cell as a macro-assisted small cell that can extend the spectrum into higher frequency bands. Multicell cooperation based on macrocell assistance of small cells aims to make efficient use of higher and wider frequency bands for small cells. In the phantom cell solution, the protocols for the control (C)-plane and the user (U)-plane are separated. Whereas the C-plane

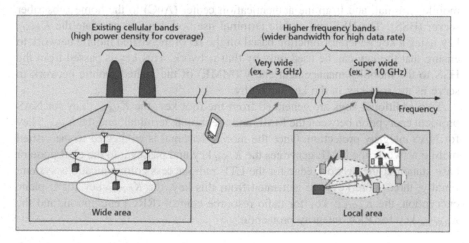

Fig. 2.7 Wide area and local area bandwidth utilization [9]

is used to transmit control signals, the U-plane is used to transmit user data. In the case of wide area deployments of macrocells, both the C-plane and the U-plane for a mobile terminal are provided by the serving macrocell as in conventional mobile communication systems. In the case of local area deployments of small cells, however, the C-plane for a mobile terminal is provided by a macrocell in a lower frequency band, while the U-plane for the mobile terminal is provided by a small cell in a higher frequency band. For this reason, macro-assisted small cells are called phantom cells because they transmit only UE-specific signals. This scheme reduces the amount of control signaling or communication overhead involved in frequent handovers between small cells on one hand, and between small cells and macrocells on the other. Thus, using small cells in higher frequency bands helps to ensure network connectivity and availability.

3G mobile networks currently provide protection for data confidentiality, authentication, C-plane and U-plane confidentiality, and C-plane integrity. The LTE next generation mobile communications system will require additional security functions, including a key hierarchy, separate security functions for access stratum (AS) and non-access stratum (NAS), and expanded forward security functions to protect handovers. The AS and NAS refer to different functional layers in the universal mobile telecommunications system (UMTS) protocol stack. Whereas the NAS specifies communication between a mobile terminal and a node in the core network, the AS specifies communication between a mobile terminal and an eNB node or base station at the network edge for the LTE radio access communications system.

Zugenmaier and Aono [10] have proposed a handover scheme that utilizes forward security to limit the scope of damage when a compromised key is used. The handover scheme is based on the key hierarchy shown in Fig. 2.8. Two keys are generated by the core network and mobile terminal (UE) during the execution of the authentication and key agreement (AKA) mechanism for mutual authentication. This includes the CK key for encryption and the IK key for integrity protection. These keys are passed from the universal subscriber identity module (USIM) to the mobile terminal, and from the authentication center (AuC) to the home subscriber server (HSS). The HSS and mobile terminal use both keys to generate the K_{ASME} key using a key generator function based on the ID of the visited mobile network to ensure that the key can be used only by that network. This key is passed from the HSS to the mobility management entity (MME) of the visited mobile network to serve as the root key in the key hierarchy.

Two additional keys are generated from the root key, the K_{NASenc} key for NAS protocol encryption between the MME and the mobile terminal, and the K_{NASint} key for NAS integrity protection. Once the mobile terminal is connected to the visited mobile network, the MME generates the K_{eNB} key and passes it to the eNB node or base station at the network edge for the LTE radio access communications system. Finally, three more keys are generated from this key, the K_{UPenc} key for U-plane encryption, the K_{RRCenc} key for radio resource control (RRC) encryption, and the K_{RRCint} key for RRC integrity protection.

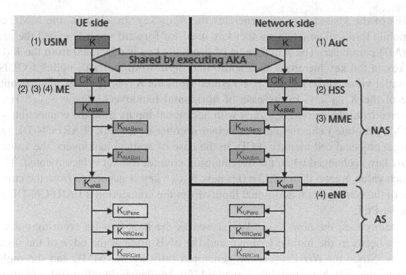

Fig. 2.8 Key hierarchy and generation in LTE [10]

The concept of forward security was introduced to provide protection from unauthorized network access in the event that an eNB node at the network edge is subverted. If an attacker gains access to an encrypted key through a compromised eNB node, forward security leverages computational complexity to ensure that the key cannot be decrypted. Figure 2.9 shows two different types of key delivery or distribution mechanisms, including horizontal and vertical handovers. When a mobile terminal is connected to an eNB node at the edge of a visited network through a shared initial AS security context, the mobility management entity

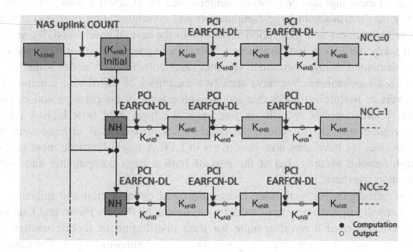

Fig. 2.9 Horizontal and vertical handovers [10]

(MME) of the visited network generates the K_{eNB} key. In addition, the MME and the mobile terminal generate another key used for forward security called the next-hop (NH) parameter. The initial value of the K_{eNB} key is generated from the K_{ASME} root key in the key hierarchy with additional input from the NAS uplink COUNT. The initial value of the NH key is generated from the K_{ASME} root key and the initial value of the K_{eNB} key. In the case of horizontal handovers, a new $K_{eNB}*$ key is generated from the current K_{eNB} key with additional inputs from the connection's E-UTRAN absolute radio frequency channel number-down link (EARFCN-DL) and its target physical cell identity (PCI). In the case of vertical handovers, the value of the NH key is changed when the NH chaining counter (NCC) is incremented. Thus, for each value greater than NCC = 0, a new $K_{eNB}*$ key is generated from the current value of the NH key with additional inputs from the connection's EARFCN-DL and its target PCI.

In both cases, the new $K_{eNB}*$ key serves as the base key for securing communication between the mobile terminal and the eNB node at the edge of the visited network. Since the NH key can be generated only by the MME and the mobile terminal, the NH keys provide a method for implementing forward security in handovers across multiple eNB nodes. Thus, at the time of vertical key delivery, the next-hop forward security mechanism ensures that the future K_{eNB} key used to connect the mobile terminal to another eNB node after n or more handovers (where $n = 1$ or 2) cannot be guessed due to computational complexity. Even if the current K_{eNB} key is leaked, threats to the network are limited because future keys are generated without using the current K_{eNB} key. Unlike horizontal key distribution, however, vertical key distribution has a large computation overhead which can delay handovers. Thus, like other security mechanisms we have discussed, forward security involves a tradeoff between enhanced security and availability.

Our discussion of trust models for secure routing in wireless sensor networks and key management schemes for secure key derivation and distribution in non-cellular wireless and mobile networks underscores the following point. Traditional security mechanisms with large computation and communication overhead are not feasible for wireless next generation networks. In the case of wireless NGNs, secure routing, key derivation and distribution, and data aggregation and storage need to be implemented using distributed schemes and collaborative nodes rather than centralized mechanisms. We have seen two examples of distributed schemes for handovers in mobile networks that require some form of cooperation and collaboration between nodes or cells. In one case, we have seen how LTE-A might employ phantom cells to reduce the communication overhead in handovers. In another case, we have seen how handovers in LTE-A might be made more secure through forward security, but at the cost of both a large computation and communication overhead.

New paradigms for distributed cooperation and collaboration and information diffusion will require a new trust model for wireless NGNs. Di Pietro and Guarino [11] have proposed a novel scheme for data distribution in mobile unattended WSNs that might be applicable to key management schemes in LTE-A. The scheme is motivated by epidemic theory. Analogous to an infected person, a node in a network can leak data if it has been infected by a malicious entity or compromised by malfunctioning behavior. Epidemic studies on how healing and infection rates affect the probability that a disease will become extinct or spread can be used to determine the replication rate necessary to prevent data leakage. Although replication and distributed dissemination provide the most straightforward approach to reliable data storage in WSNs, this approach compromises confidentiality and source location privacy. Similarly, although key derivation and distribution provide the most straightforward approach to handovers, this approach compromises the same cybersecurity principles since keys can be intercepted in the handover process and subsequently decrypted. Instead, Di Pietro and Guarino argue that local information sharing schemes are the best solution to security in a mobile context.

To implement a local information sharing scheme, a WSN is used to monitor a specified area with a variable number of sensors depending on the requirements of the security domain. After acquiring data, sensor nodes collaborate to securely route the data to one or more sinks for storage. The network topology helps to ensure availability by reducing data leakage due to the malicious capture or failure of a sensor node. The network topology helps to ensure confidentiality by preventing exposure of stored data such as secret keys to unauthorized entities. Instead of protecting individual pieces of information sensed by each node in the network, the network topology protects the information sensed by the network as a whole. By making sure that the pieces of information stored by the sensor nodes in the network are uncorrelated with their location, the quantity and content of information that can be recovered or inferred from a given fraction of the network depends on the size of the fraction rather than on the identity of the nodes. To limit the bandwidth requirements and energy consumption of sensor nodes, the sensed data needs to be locally transmitted. Here mobility, which is often seen as a liability in a security context, is exploited as an asset to spatially diffuse or spread the sensed data throughout the monitored area of the network in an energy efficient fashion.

This can be accomplished by establishing a local information sharing scheme over the monitored area, where one share of the secret information is randomly distributed to each of k neighbor nodes of the distributing node. The property of mobility is then leveraged to randomly move these nodes around until the secret information is spatially spread out over the localized area. The shares of secret information are diffused over a localized area through the mobility of sensor movements. Subsequently, a sink can be used to explore the monitored area to retrieve all the shares of information stored by the nodes within the local communication range of the sink. Such a local information sharing scheme provides availability, which is the main benefit of data replication, without the cost of potential data leakage and loss of confidentiality. The primary goal of confidentiality is enforced by requiring access to at least k nodes in the network to recover a single piece of secret information. Information diffusion effectively renders the content of the data stored in a node independent of its location, thereby providing both source and location confidentiality. The remaining chapters will develop a virtue-based trust model that can be used to support the cooperation and collaboration required between network entities in local information sharing schemes.

References

1. Kisner RA et al (2010) Cybersecurity through real-time distributed control systems. Oak Ridge National Laboratory, ORNL/TM-2010/30, pp 1–14
2. Vasilache RA et al (2013) Comparative evaluation of trust mechanisms for wireless sensor networks. In: 11th RoEduNet international conference, Sinaia, Romania, 17–19 Jan 2013, pp 1–5
3. Barker E et al (2013) A framework for designing cryptographic key management systems. NIST, Special Publication 800–130. http://nvlpubs.nist.gov/nistpubs/SpecialPublications/NIST.SP.800-130.pdf. Accessed 15 Jul 2014
4. Barker E et al (2012) Recommendations for key management. NIST, Special Publication 800–57, Part 1, rev 3. http://csrc.nist.gov/publications/nistpubs/800-57/sp800-57_part1_rev3_general.pdf. Accessed 15 Jul 2014
5. Bergstra J, Burgess M (2009) Local and global trust based on the concept of promises. Computing Research Repository (CORR), paper 0912.4637. http://arxiv.org/pdf/0912.4637.pdf. Accessed 15 Jul 2014
6. Hoeper K et al (2008) Security challenges in seamless mobility–how to 'handover' the keys? In: 4th international ACM wireless internet conference (WICON '08), Maui, HI, 17–19 Nov 2008, pp 2–3
7. Chen L, Gong G (2012) Wireless security: security for mobility. In: Communication system security. CRC Press, New York
8. Forsberg D et al (2013) LTE security, Chap 2, 2nd edn. Wiley, Hoboken
9. Kishiyama Y et al (2013) Future steps of LTE-A: evolution toward integration of local area and wide area systems. IEEE Wirel Commun 20(1):12–18. doi:10.1109/MWC.2013.6472194
10. Zugenmaier A, Aono H (2013) Security technology for SAE/LTE (system architecture evolution 2/LTE). NTT DOCOMO Tech J 11(3):28–30. http://www.nttdocomo.co.jp/english/binary/pdf/corporate/technology/rd/technical_journal/bn/vol11_3/vol11_3_027en.pdf. Accessed 15 Jul 2014
11. Di Pietro R, Guarino S (2013) Confidentiality and availability issues in mobile unattended wireless sensor networks. In: 14th IEEE international symposium and workshops on a world of wireless, mobile and multimedia networks (WoWMoM), Madrid, Spain, 4–7 Jun 2013, pp 1–6

Chapter 3
Trust, Epistemic Normativity, and Rationality

Social scientists have criticized epistemology, or the theory of knowledge, for its narrow emphasis on epistemic rationality or normativity. They argue that contemporary epistemology is obsessed with the normative question of what justification conditions a true belief needs to qualify as knowledge. In particular, whereas epistemic normativity is based on the synchronic intellectual goal of now having accurate and more comprehensively coherent beliefs, they argue that the rational assessments required in everyday activities in business, economics, and politics involve diverse goals, pragmatic as well as intellectual, and that these goals may be long-term as well as short-term. Perhaps most importantly, they argue that there are practical limitations on how much time and effort it is reasonable to spend pursuing the intellectual goal of attaining more accurate and comprehensively coherent beliefs. Computer networks, for example, often rely on autonomous rational agents to make apt decisions about which entities to trust based on trust models, mechanisms, and methods that must operate within time and space complexity constraints. Thus, social scientists argue that what is needed is not a general theory of knowledge, but a general theory of rationality that provides norms for evaluating the success of our cognitive performances in everyday rational activities. Such a general theory should assist us in judging the success or reliability of mechanisms, both human and machine, as they seek to achieve their pragmatic goals.

When we turn to the question of rationality, however, we encounter an ineliminable circularity. Trust and rationality seem viciously circular. We cannot exercise our rational faculties unless we trust their reliability, but we cannot trust their reliability unless we confirm the reliability of their operation and deliverances using those very faculties. Foley [1] argues that a general theory of rationality can be developed in terms of the more restrictive notion of epistemic rationality or normativity which avoids vicious circularity. According to Foley, the notion of epistemic rationality has two theoretical advantages over other theories of rationality. First, the notion has been rigorously developed in several different conceptions of knowledge such as foundationalism, coherentism, and reliabilism. Secondly, the notion can be used to develop a non-circular theory of rationality by seating some of its key ideas in a social context. Regardless of which account of rational belief is offered, the account can escape vicious circularity because it makes

© The Author(s) 2014
M.G. Harvey, *Wireless Next Generation Networks*,
SpringerBriefs in Electrical and Computer Engineering,
DOI 10.1007/978-3-319-11903-8_3

no reference to any other notion of rationality than epistemic rationality. We will return to Foley's argument after considering an account of epistemic rationality or normativity that seems well suited to this role.

3.1 Motivations for Virtue Perspectivism

Sosa [2, 3] develops an account of epistemic rationality or normativity which broadens our conception of rationality by including trust as an epistemic mechanism, disposition, or competence that is just as foundational to rationality as reason with its associated intellectual virtues. According to Sosa's virtue perspectivism, trust and rationality reflect two different but mutually dependent epistemic levels, which allegedly interact in a virtuous circle to achieve the intellectual goal of attaining a more accurate and comprehensively coherent perspective on the basis of which we can evaluate both our explicit and implicit beliefs or commitments. This account of epistemic normativity might not only provide the theoretical anchor for other theories and notions of rationality. It might provide the vital link needed to relate trust and rationality in a non-circular theory of rationality that can be translated into a normative, virtue-based trust model for wireless next generation networks. Since virtue and character play a central role in most societies, moreover, a virtue-based trust model can be adapted to a variety of cultures which may differ on the relative importance placed on different intellectual virtues.

In rough outline, Sosa argues that belief acquisition and belief revision may be best understood as cognitive competences, intellectual virtues, or skills that are exercised by a subject at two different epistemic levels. Whereas the first type of competence is exercised at a lower epistemic level called animal knowledge, the second type of competence is exercised at a higher epistemic level called reflective knowledge. Animal knowledge is unreflective knowledge comprised of first-order commitments or beliefs that are automatically produced by a reliable belief-forming disposition that is part of our natural endowment. In contrast, reflective knowledge is comprised of second-order commitments or meta-beliefs about the sources of our first-order commitments or beliefs and their reliability. Whereas first-order beliefs are acquired unconsciously at the level of animal knowledge, meta-beliefs are acquired consciously at the level of reflective knowledge by monitoring the reliability of our ground-level belief-forming and belief-sustaining competences. The distinctive value of knowledge, in contrast to mere true belief, lies in the development of a conscious evaluation or epistemic perspective on one's largely unconscious cognitive doings.

Since Plato, philosophers have agreed that a belief needs to be true to qualify as knowledge. The question that has divided philosophers to the present is whether this requirement is sufficient for knowledge. For years, responses to the question of what justification conditions a true belief needs to qualify as knowledge remained divided between two rivals in epistemology—coherentism and classical foundationalism. Like other contemporary epistemologists, Sosa [4] has described these two different theories of knowledge with the metaphors of the "raft" and the "pyramid" respectively. According to foundationalism, there must be ultimate premises, bases, or sources of knowledge that get at least some of their justification by means other than reasoning from further premises. Foundational beliefs must be justified non-inferentially to stop the potential infinite regress of still more premises.

For Sosa, the foundationalist's claim that experientially given facts directly justify at least some of our beliefs merely through their truth is epistemically inadequate for two reasons. First, such foundational beliefs lack sufficient content to explain the richness of our knowledge of an external world. Secondly, and perhaps most importantly, when the foundationalist claims that we know something merely because it is true, this explanation does not rise to the level of reflective knowledge. Although the explanation is appealing because it is both necessary and epistemically simple, it is unilluminating because it does not address the deeper epistemic question of whether the mere fact that a belief is true is sufficient to justify our believing it. Since the foundationalist ignores this question, Sosa argues that he is led to cognitive quietism, whereby he does not seek an account of epistemic normativity that is more epistemically rewarding.

According to coherentism, the fact that a belief is part of a more comprehensively coherent system of belief that is truth-conducive is sufficient to justify our believing it. As Sosa points out, however, epistemology is filled with mythological creatures such as the genius mad man or the brain in a vat, both of whom may have comprehensively coherent pictures of the world but who nevertheless remain out of touch with truth. Analogous to the infinite regress in foundationalism, moreover, coherentism is vexed by the problem of epistemic circularity. That beliefs derive at least some of their justification from mutually supportive basing relations with other beliefs presupposes that the belief being used as a premise is itself justified. Thus, Sosa [3, p. 87] argues that foundationalism and coherentism ultimately face the same problem. Inferential reasoning transfers justification from premises to conclusions, but in both theories of knowledge this presupposes that the premises are already justified. Neither the pure regress in foundationalism nor the pure circle in coherentism is able to explain this problem.

Sosa's virtue perspectivism is a species of virtue reliabilism that attempts to combine the strengths of foundationalism and coherentism while avoiding their weaknesses. Virtue reliabilism is in turn a species of reliabilism as well as of virtue epistemology. According to reliabilism, a true belief qualifies as knowledge or is justified if it arises from a reliable faculty or process for attaining truth. Zagzebski and Fairweather [5] further argue that reliabilism is itself a species of externalism, which holds that the conditions for justification need not be accessible to one's consciousness as in internalism. Since the conditions for justification can be external, they point out that reliabilism can also be considered a species of naturalized epistemology, which holds that normative epistemic properties such as justification can be reduced to natural, non-epistemic properties.

Virtue epistemology is distinguished from other theories of knowledge by its emphasis on the subject rather than his or her belief as the seat of justification. Thus, according to virtue reliabilism, Sosa [3, p. 135] argues that a true belief qualifies as knowledge only if its correctness derives from its manifesting certain cognitive virtues of the subject, where a cognitive virtue is defined as a truth-conducive disposition. Sosa's claim involves a criticism of internalism associated with foundationalist theories of knowledge, where a belief qualifies as knowledge only if it is backed with reasons or arguments adduced as premises and the subject is aware or conscious of these premises at the time the belief is held. For this reason, virtue reliabilism is described as a species of externalism. Sosa's [3, p. 191] externalist theory of knowledge aims to combine internal coherence with externally competent faculties that do not depend on our awareness for their correct operation, which together give us a more accurate and comprehensively coherent grasp of truth.

Sosa's variant of virtue reliabilism is distinguished from other reliabilist theories of knowledge in its emphasis on the subject as the seat of justification and his or her cognitive virtues or aptitudes. There are several different species of reliabilist theories of knowledge. Tracking accounts focus on a belief's counterfactual relation to the truth of what is believed; reliable indicator accounts focus on the properties of a belief itself or the reasons for holding it which must be sufficient for its truth; reliable process accounts focus on the cognitive process or mechanism that forms the belief and on the truth ratio in the products of that process, actual and counterfactual. In contrast to these other reliabilist accounts of knowledge, as well as to coherentism, Sosa's [3, p. 187] virtue perspectivism claims that in order to qualify as knowledge the correctness of a belief must derive from a competence exercised in appropriate conditions, epistemically so, in a sense that goes beyond its being just a belief that coheres well within the subject's epistemic perspective.

Virtue perspectivism is based on a novel, albeit controversial, interpretation of René Descartes' epistemological project. Far from being a simple foundationalist project as portrayed by the conventional interpretation, Sosa argues that Descartes' epistemological project involves a more subtle and complex mix of foundationalism, coherentism, and reliabilism. Accordingly, Sosa [3, pp. 139–140] argues that Descartes sought to establish that our beliefs derive from sources reliably worthy of our trust. In order to place our scientific knowledge on a firm footing, Descartes considers skeptical doubts that target his intellectual faculties of perception, memory, introspection and intuitive reason, whereby one might know such things as $3 + 2 = 5$ or that if one thinks one exists. Sosa argues that, for Descartes, these skeptical doubts can be defended against only by coherence-inducing theological reasoning that yields an epistemic perspective on himself and his world on the basis of which he might confidently trust the reliability of his faculties.

Sosa [3, p. 141] argues that the problem of epistemic circularity arises because the faculties whose reliability he seeks to endorse include the very faculties he must rely on in arriving at such an endorsing perspective. Sosa argues, moreover, that Descartes himself provides a way out of this circle. Although one must underwrite the reliability of his faculties at a later stage, Sosa [3, p. 151] tells us that what enables one to do so is the epistemically appropriate use of those faculties in yielding the required epistemic perspective. According to Sosa [2, pp. 131–132], Descartes' epistemological project is best understood as involving two distinct stages. In the first stage, Descartes meditates about the issues raised by his skeptical doubts with the same sort of epistemic justification and even certainty that characterizes the reasoning of an atheist mathematician. These meditations, however, are deprived of an epistemic perspective or worldview by which they may be seen as epistemically propitious. Absent a proper epistemic perspective, such reasoning cannot rise above the level of *cognitio*. And yet, over time, enough pieces of this reasoning may come together through abduction into a view of ourselves and our place in the scheme of things that is sufficiently comprehensively coherent to raise us above that level into a higher, more reflective and enlightened knowledge called *scientia*.

Descartes' distinction between *cognitio* and *scientia* motivates Sosa's distinction between animal knowledge and reflective knowledge. *Cognitio*, or animal knowledge, requires that one attain the truth of a matter by being appropriately constituted by nature, as well as appropriately situated in the environment, to issue reliable judgments about how things stand. Unlike Descartes' conception of reliability which requires infallible reliability, however, Sosa's [3, p. 199] conception requires only a high degree of reliability. For Sosa, reliability is a matter of degree which depends on how easily one might go wrong in thinking as one does through exercising the relevant dispositions, faculties, or intellectual virtues. *Cognitio*, moreover, requires that the correctness of one's belief be attributable to the exercise of such a competence in its appropriate conditions. In contrast, *scientia* or reflective knowledge requires an adequate epistemic perspective on one's cognitive doings. Sosa [2, p. 130] argues that this higher state of knowledge is attained only if one can see how one is acquiring and sustaining the belief whose source is at issue. *Scientia*, moreover, requires that one see that way as reliable, as one that tends to lead one aright rather than astray.

3.2 Animal Knowledge and Reflective Knowledge

Thus motivated by Descartes' distinction between *cognitio* and *scientia*, Sosa draws a subtle and deep distinction between two types of rational justification, unreflective and reflective, based on his distinction between animal knowledge and reflective knowledge. Unreflective rational justification does not require reflective endorsement of either the specific or generic reliability of one's sources of knowledge. In contrast, reflective rational justification, or epistemic justification properly understood, is acquired partly through reason-based competences, skills, aptitudes, or intellectual virtues such as open-mindedness, carefulness, understanding and interpretation, and partly through an endorsing perspective on the reliability of our animal competences which are not reason-based but whose deliverances are trustworthy because the source is reliable. Reflective rational justification and unreflective rational justification involve two distinct types of competences, the first reason-based and the second not reason-based but animal. Sosa [2, p. 87] argues that this theoretical distinction may be seen empirically in the fact that a belief-forming process or mechanism can be something close to a reflex, or it can be a very high-order, central-processing ability that involves the weighing of pros and cons.

The distinction between reason-based competences at the level of reflective knowledge and competences that are not reason-based at the level of animal knowledge, and the corresponding distinction between reflective rational justification and unreflective rational justification, is meant to broaden the scope of the conventional definition of rationality as involving only reason-based competences and reflective rational justification. Both the concept of rationality and the cognate concept of rational justification are thereby enlarged and illuminated within an epistemic perspective. This broadening of the scope of the conventional definition of rationality has two implications. First, rational justification involves not only internal competences such as consciousness and the intellectual virtues that depend on it. More fundamentally, rational justification involves external competences that do not depend for their reliable operation on being conscious or aware of their operation. Secondly, in contrast to reflective knowledge, animal knowledge is not irrational but corresponds to a foundational and minimal type of rationality that lacks the reflexivity normally associated with being rational. At the level of animal knowledge, one can be rational without being rational reflectively or self-consciously so, though perhaps not in the fullest sense of rationality as defined by reflective knowledge according to the internalist requirement of a conscious, endorsing perspective.

Thus, Sosa argues that animal competence is a distinctive type of epistemic standing that is attained without the aid of an endorsing perspective, whereby the competence is seen as reliable. Animal competences must have their own epistemic standing, since the epistemic standing of one's explicit or implicit beliefs or commitments cannot be generated entirely through reasoning based on beliefs acquired only through unjustified or epistemically inappropriate commitments. Unlike the epistemic standing of reason-based competences, the epistemic standing of animal competences does not depend on any justificatory performance by their owner. Instead, it derives from their animal reliability in enabling us to harvest needful information about the world, and from their being part of the animal endowment of an epistemically proper-functioning human being.

Sosa [3, pp. 237–238] argues that our trust in our animal faculties such as perception and memory is a source of epistemic standing for the beliefs acquired through them. In contrast to reason-based competences such as intellectual virtues, Sosa suggests that we can think of trust as a distinctive type of animal competence or disposition that is not reason-based but which nevertheless enables us to trust our other animal competences or faculties such as perception. Thus, both trust and reason may be seen as valid sources of epistemic standing for our beliefs. Moreover, since trust can be regarded as an animal competence for reliably acquiring or forming beliefs, and since animal knowledge is not irrational, it follows that trust is not irrational. Instead, trust is an integral part of our rationality. Indeed, trust is the very foundation of our rational nature.

Although Sosa [3, pp. 235–236] acknowledges that he may be stretching the term "competence" or the general notion of an intellectual virtue in the case of animal competences that are not reason-based—perhaps because they do not involve consciousness and deliberation as in the case of reason-based competences typically associated with intellectual virtues—the more general conception of a competence or intellectual virtue allows us to assign these belief-forming and belief-sustaining dispositions normative status. In order to qualify as knowledge, even in the lower sense of animal knowledge, these dispositions need to be epistemically evaluable. Something must explain their epistemic standing, namely, the truth-reliability of the dispositions.

The exercise of a competence, whether reason-based or not, can be apt or deficient in its performance. Beliefs can be well formed or not, both at the lower level of animal knowledge and at the higher level of reflective knowledge. Thus, in response to the value problem posed in Plato's *Meno* concerning what makes knowledge better than mere true belief, Sosa [2, p. 93] answers that it is apt belief. According to virtue epistemology, the value of apt belief is as epistemically fundamental as true belief. Epistemic or intellectual virtues contribute to value not just instrumentally but constitutively. Intellectual virtues are constitutive of knowledge because the aptness of a belief consists in its being accurate because competent. Beliefs can be thought of as a special case of performances, namely, epistemic or cognitive performances. In the case of animal knowledge, where higher, reason-based intellectual virtues are not involved, a belief acquired or formed through an animal competence is apt if its correctness or accuracy is attributable to the exercise of this competence by the subject in its appropriate conditions rather than to brute luck. The evaluation of a belief in terms of a cognitive performance, Sosa [3, p. 80] argues, is no more strange than the evaluation of an archer's shot as skillful, which remains an evaluation of the shot itself even if it depends on factors involving the surroundings.

These considerations lead to Sosa's [2, p. 109] definition of animal knowledge as apt belief, whereby one knows that proposition *p*.

Kp For any correct belief (or presupposition) that *p*, its correctness is attributable to a competence only if it derives from the exercise of that competence in conditions appropriate for its exercise, where that exercise in those conditions would not too easily have issued a false belief (or presupposition).

The requirement of apt belief for knowledge is meant to explain how someone can know something without reasoning from prior knowledge, that is, non-inferentially. According to Sosa [3, p. 186], one can know something non-inferentially so long as it is no accident or coincidence that one is right. To qualify as knowledge, a belief must not only be accurate or true as in foundationalism, it must also be apt in the way it is acquired and sustained.

For Sosa [2, p. 29], the aptness of a belief is not a property of the belief, but a property of the subject in the manifestation of a disposition or competence, one which exercised in appropriately normal conditions ensures, or at least makes highly probable, the success of any performance issued by it. A belief is apt if it is acquired through the exercise of a ground-level, animal competence. When we trust

our color vision in taking the surface of a wall to be red, for example, Sosa
[2, pp. 32–33] tells us that the competence we exercise seems to be a sort of default
competence, whereby one automatically takes the light to be normal absent any
indication to the contrary. Apt belief is characterized by an automatic, spontaneous,
instinctive reflex or reaction to trust the reliability of our animal faculties such as
perception.

In general, Sosa argues that an epistemic competence, whether reason-based or
not, is a disposition to trust or accept the deliverances of an epistemic source, where
the source is itself a disposition to host certain intuitions or intellectual seemings.
Sosa [2, p. 106] also argues that animal knowledge is one sort of epistemic competence, namely, the disposition to implicitly trust a source of knowledge absent
any indication to the contrary. It is not entirely clear, however, whether trust is itself
an animal competence or whether it is an accompanying feature of an animal
competence such as perception. If trust is itself an animal competence, other animal
competences such as perception and memory would seem to depend on it, making
trust more basic. Taking this ambiguity into consideration, Sosa seems to suggest
that trust is itself an epistemic source, disposition, or animal competence that
enables us to accept the deliverances of other ground-level belief-forming and
belief-sustaining animal dispositions such as perception. This interpretation is
supported elsewhere where Sosa [3, p. 237] tells us that cognitive scientists suggest
that the reliability of our implicit commitment to repeatedly take visual experience
at face value is itself delivered by a competence that is not reason-based. This
epistemic source, disposition, or animal competence is justified because it is reliable; the animal beliefs or commitments immediately delivered by the faculties that
depend on it are preponderantly true, even if occasionally false. Although animal
competences operate automatically on an unconscious level, Sosa [2, p. 60] argues
that we can still think of them as intellectual competences because they system-
atically lead us aright in our believings.

It is important to stress the unreflective or non-rational status of these animal
competences. We are disposed by our constitution as rational creatures to sponta-
neously form perceptual beliefs about the world and to store those that we need in
memory. According to Sosa [3, p. 243], these unconscious dispositions, perhaps
innate and triggered through normal infancy, receive little or no benefit of conscious
rational support or development in their early stages. Of course, this situation
changes in later stages as we become more rationally complete beings in fuller
conscious control of the beliefs we acquire and sustain. Although animal knowledge may involve reasoning in the unreflective sense of inferential patterns or habits
of thought which cognitive psychologists call procedural knowledge, Sosa argues
that these patterns or habits are not consciously and deliberately chosen or deduced
by the subject. Nevertheless, in order to avoid cognitive quietism, Sosa argues that
we can think of animal competences as involving implicit states that are epistemically evaluable. Whether we call these implicit states beliefs or not, they are
states of implicit adherence to principles or commitments that one holds.

In contrast to animal knowledge, which has to do with the acquisition and
sustenance of apt belief, Sosa [3, p. 75] argues that reflective knowledge requires

that the beliefs thus formed be placed in an epistemic perspective within which they may be seen as apt. This requirement can be expressed more formally, $K^+p \leftrightarrow KKp$, where K is animal knowledge and K^+ is reflective knowledge. Reflective knowledge that p requires defensibly apt belief, that is, meta-belief whereby the subject aptly believes that a belief has been aptly acquired and sustained. Thus, reason-based competences or intellectual virtues such as open-mindedness, carefulness, understanding, and interpretation are required to monitor the reliability of our animal competences.

The distinction between animal knowledge that p (Kp) and reflective knowledge that p (K^+p) is motivated by the following problematic. Our animal knowledge is comprised of beliefs that are groundless, that is, not reason-based. Animal beliefs are held without any supporting reasons or arguments. Whereas some of these beliefs qualify as knowledge of the obvious, others are superstitions or dogmas. Thus, Sosa [3, p. 163] argues that such a distinction is needed to explain the difference, epistemically, between these two sorts of animal beliefs. The distinction is also motivated by the ethics of belief, a deontological approach to epistemology analogous to ethics, where we have a duty or responsibility to hold beliefs that are better justified. This approach claims that it is better to believe and act in ways that are reflectively right than in ways that just happen to be right but unreflectively so. Similarly, Sosa [2, p. 129] argues that it is more admirable to attain one's intellectual goal of having more accurate and comprehensively coherent beliefs through one's own thought and efforts than it is to be a passive recipient of brute luck.

The distinction between animal knowledge and reflective knowledge, moreover, reflects the fact that knowledge is a matter of degree. One knows some things better than other things. Furthermore, people who are experts in a field know things in that field better than people who are not experts in the field. Thus, epistemic rationality or normativity has to do with the justification, degree, or quality of one's knowledge. Sosa [3, p. 136] argues that a belief is of higher epistemic quality if it is safer, one that might not too easily have been wrong, or one that is better justified where one's evidence for the belief is stronger as in foundationalism or the belief has been more reliably acquired and sustained as in virtue reliabilism.

The continuum or gradient in the degree or quality of one's knowledge extends from lower animals and human infants to more rationally complete adult humans. According to Sosa [3, p. 63], the knowledge of lower animals and infants, as well as some mature human knowledge, has little reflection. In contrast, adult humans have a rich epistemic perspective, even if many of the beliefs or commitments it entails remain implicit and mostly beyond our capacity to articulate. Sosa [3, p. 147] argues that reflective knowledge is a higher state of knowledge that requires a view of ourselves, beliefs, faculties, and situation in the light of which we can see the sources of our beliefs as sufficiently reliable. Thus, knowledge is most adequately expressed in an epistemic perspective or worldview as defined by the principle of the criterion.

PC   Knowledge is enhanced through justified trust in the reliability of its sources.

This principle states that the epistemic quality of a belief rises with justified awareness of the reliability of one's sources. Although the PC principle is defined for explicit beliefs or commitments formed by reason-based competences such as intellectual virtues, Sosa [3, p. 139] argues that the principle can be extended to implicit beliefs or commitments formed by animal competences so long as these beliefs are able to rise to the level of reflective knowledge by meeting the requirement that one must be able to defend the reliability of one's sources. Thus, the requirement for reflective knowledge stated by the PC principle requires in turn that the implicit beliefs or commitments that comprise our animal knowledge be normative, that is, epistemically evaluable.

To satisfy the PC principle, the relation between animal knowledge and reflective knowledge must involve some sort of interaction that enables implicit beliefs or commitments formed by animal competences to rise to the higher level of explicit beliefs or commitments formed by intellectual virtues. The question is how such a transmutation of mere true belief at the level of animal knowledge into reflective knowledge is possible. That some sort of interaction between these two levels of knowledge and classes of beliefs is required implies a more subtle and complex relation between reason-based competences that yield justified trust in the reliability of our sources of knowledge and the animal competence that makes trust itself possible.

Accordingly, Sosa suggests that the relation between internalism and externalism is not one of mutual exclusivity, but one which requires a more nuanced understanding of their complementarity. Reason-based competences and animal competences involve internalist and externalist components respectively that depend on each other and must cooperate in the transmutation of animal belief into reflective belief. These two different types of competences, with their respective internalist and externalist components, are not mutually exclusive alternatives we must choose between in our epistemology, one rational and the other irrational. Instead, they are best understood as mutually dependent and complementary cognitive processes or mechanisms that are intended by nature to work together in making us more complete rational beings.

Whereas animal knowledge is justified externally through reliable, trustworthy competences that are not reason-based, reflective knowledge is justified internally through reason-based competences such as intellectual virtues. In animal knowledge, justification derives from a source other than the subject's reasoning, namely, basic trust. Thus, for Sosa [3, p. 135], the source of justification for a belief need not require consciousness as in internalism. In reflective knowledge, justification derives from a conscious and deliberate evaluation of how well a performance issued by an animal competence puts us in touch with truth. This evaluation requires that the subject have an epistemic perspective on his belief, whereby he is able to endorse or underwrite the source of the belief as reliably truth-conducive.

The distinction between animal knowledge and reflective knowledge allows us to explain the rich content of human knowledge. The distinction reflects the fact that our cognitive structure, epistemic perspective, or worldview is comprised of two distinct classes of beliefs or commitments, one implicit and mostly inarticulable and the other explicit and fully articulable. Thus, Sosa argues that a person may simultaneously hold fully articulable scientific or theological beliefs and implicit and inarticulable beliefs. At any given moment, our cognitive structure is mixed with elements that vary variously. What is important is that these implicit states be epistemically evaluable in at least some of the same ways as explicit states, whatever their epistemic status and causal nature. For Sosa, epistemic evaluation includes both a dimension of aptness and a dimension of rational justification.

The meta-perspectival component of any cognitive structure is comprised of meta-beliefs, that is, beliefs about the body of beliefs one has acquired and sustained through the unconscious and unreflective mechanisms or competences of animal knowledge. Sosa [3, p. 66] argues that this perspective allows us to take a view, at least partially, of those body of beliefs. Using our various faculties and subfaculties, including those constituted by unconscious inferential patterns or habits of thought, we may reflectively attain a worldview that is fundamentally different from the largely implicit prior knowledge or framework of beliefs that precedes it. Sosa [3, p. 77] argues that this new framework of beliefs is distinguished from the previous framework by three epistemically significant features. First, our previous implicit beliefs or commitments are now explicit through an enlightened perspective. Secondly, such consciousness-raising enables greater comprehensive coherence, whereby our beliefs are more tightly integrated and visibly interrelated with other beliefs. Thirdly, from the vantage point of such a reflective perspective, we can moderate the epistemic influences of our unconscious cognitive habits on the formation of our implicit beliefs or commitments and change these beliefs.

3.3 Epistemic Circularity and Cross-Level Coherence

Virtue perspectivism is an example of experimental philosophy that aims to illuminate issues in epistemology such as epistemic normativity with empirical research from cognitive science. The theory tries to integrate conflicting elements from foundationalism, coherentism, reliabilism, internalism, and externalism in the light of such research to avoid cognitive quietism and advance discussions in epistemology beyond current impasses. The conception of animal knowledge as apt belief Kp, which involves a distinctive type of competence that has its own justification without being reason-based, is meant to stop the infinite regress in foundationalism. The conception of an endorsing perspective K^+p, which allegedly satisfies the PC principle, is meant to avoid vicious circularity.

In his response both to the pure regress and to the pure circle, Sosa argues that skepticism is best understood as an attack against reflective knowledge rather than animal knowledge. Specifically, skepticism is an attack against the possibility of satisfying the requirement of an endorsing perspective as defined by the PC principle in the light of which we can justifiedly know the reliability of our sources of knowledge. Thus, when we reply to the skeptic as Peter Strawson does that animal knowledge requires no reflective status and that we are content with our claims to such knowledge, we dodge the skeptic's challenge which leads to cognitive quietism. So long as we believe that reflective knowledge is a higher and more desirable state of knowledge, however, Sosa [2, p. 116] rightly argues that the skeptic's challenge to knowledge and the reliability of its sources will retain its interest. Thus, we must find a way to satisfactorily meet the skeptic's challenge, which is defined by the key requirement that any reliabilist theory of knowledge must meet according to internalism.

KR   A potential knowledge source K can yield knowledge for subject S only if S knows that K is reliable.

This internalist requirement for a legitimating account of justification states that one must establish the reliability of a cognitive process, mechanism, or competence before one is justified in trusting its deliverances.

Sosa's response to the KR requirement involves four stages. In the first stage of his response, Sosa [3, p. 216] argues that the internalist requirement for a legitimating account of justification is too narrowly defined. It is limited to explicit beliefs or commitments which comprise only a small fraction of our total cognitive structure. According to internalism, when a belief is based on reasons and is motivated by them, one tends to have access to the fact that those are your reasons. Here, in one sense, the source of one's knowledge is the rational basis of the belief. But in another sense, the source of one's knowledge is the source of epistemic justification for the belief, which is the belief's having the rational basis that is in the first sense its source. Sosa [3, p. 217] observes that most of one's beliefs, however, consist of implicit beliefs or commitments that are not held for sufficient reasons that are motivationally operative at the very moment when they are held. The rational aetiology of these beliefs, moreover, is now beyond our ability to recall. Since we now lack access to the sources of these beliefs, it is difficult to see how we could presently know their sources to be reliable.

The narrow scope of the KR requirement motivates the need for a more general requirement for justification that includes both types of beliefs or commitments in our cognitive structure, implicit as well as explicit. There must be a way in which some beliefs get their justification externally, not just internally. This intuition motivates the distinction between beliefs or commitments that are reason-based as in internalism, and beliefs or commitments that are not reason-based as in externalism. Sosa [3, p. 212] argues that a "source" of belief can be more broadly interpreted as the "basis" for holding the belief. Furthermore, unlike internalism, we need not think of a source as necessarily involving a rational basis. Instead, we might appeal to the more general conception of an epistemic competence. Then we may distinguish between reason-based competences, which are reliable dispositions to believe correctly based on rational inputs to which we have access as in internalism, and animal competences, which are also reliable dispositions to believe correctly but without being based on reasons as in externalism. If we interpret the KR requirement this way, the sources of knowledge whose reliability is at issue would be competences of the believer's. Whereas animal competences involve belief-forming and belief-sustaining dispositions, Sosa [3, pp. 214–215] argues that reason-based competences involve belief-monitoring and belief-revision dispositions such as intellectual virtues, properly speaking.

In the second stage of his response to the KR requirement, Sosa draws a distinction between hierarchically basic knowledge and inferentially basic knowledge. In the case of hierarchically basic knowledge, knowing that p is hierarchically basic if and only if the belief is held unaccompanied by knowledge that its source or basis is reliable. In the case of inferentially basic knowledge, knowing that p is inferentially basic if and only if the belief is not based on inference so that its source or basis is non-inferential. Furthermore, hierarchically basic knowledge may or may not include inferentially basic knowledge. This distinction allows us to include some minimal forms of reasoning in hierarchically basic knowledge such as inferential patterns or habits of thought that are not consciously and deliberately arrived at through inference. Thus, hierarchically basic knowledge is distinguished by the lack of awareness of the reliability of its sources as required by internalism, and may or may not be the product of unconscious inferential patterns or habits of thought.

In the third stage of his response to the KR requirement, Sosa [3, p. 228] defines a 2-level basic knowledge structure based on the distinction between animal knowledge and reflective knowledge analogous to Cohen's [6] basic knowledge structure which contains two theses.

T1   There is hierarchically basic knowledge, where knowing that p is hierarchically basic if and only if the belief is held unaccompanied by knowledge that its source or basis is reliable (repeated from above).

T2   Hierarchically basic knowledge provides a basis for knowing that the sources of one's knowledge are reliable both in specific cases and more generically.

To avoid the infinite regress of foundationalism, reflection must give out at some level n where we no longer have an explicit or implicit belief $B^{n+1}p$ supporting our belief $B^n p$. Thus, Sosa argues that any plausible epistemology will satisfy T1.

This leaves only T2 for the skeptic to attack, which he does with the following argument that employs a proof by contradiction.

P1   If there is human knowledge at all, there is hierarchically basic knowledge (T1).

P2   If one knows that p in a hierarchically basic way, one can come to know based partly on this basic knowledge that the basis for the knowledge is specifically or generically reliable (T2).

P3   In no case where one knows that p can one come to know based even partly on this basic knowledge that the basis for the knowledge is specifically or generically reliable (not-T2).

C   Therefore, there is no human knowledge (not-T1).

If there is some human knowledge, however, the first premise P1 is indisputable. Thus, in order to refute the skeptic's *reductio* argument, we need to show that either premise P2 or P3 is false to avert the contradiction.

In the final stage of his response to the KR requirement, Sosa tries to show that P3 is false. Sosa begins by considering Thomas Reid's solution based on the principles of Common Sense, which are comprised of presuppositions, implicit beliefs or commitments that guide our rational conduct. As such, these principles are constitutive of epistemic normativity. In order to construct arguments for the justification of our implicit belief in or commitment to these principles which seem irresistible, we need to exercise our perception, memory, and reason. Sosa [3, p. 62] argues that it is difficult to see how one can arrive at the conclusion that such faculties are reliable, however, unless one already presupposes that they are reliable in that very reasoning. Such an approach seems viciously circular. Like Descartes, Reid argues that the reliability of our faculties is guaranteed by God because he is no deceiver. But this *a priori* theological solution to vicious circularity is also viciously circular. Both Descartes and Reid, Sosa argues, presuppose that the very faculties that enable us to draw this theological conclusion are already guaranteed. Such an *a priori* approach does not tell us how one can justifiedly believe that her faculties are reliable without presupposing that they are reliable.

Nevertheless, Sosa argues that Descartes and Reid suggest a way out of the circle based on a distinction between unreflective knowledge and reflective knowledge. One cannot reach the Cartesian or Reidian worldview using faculties that include claims about the reliability of those faculties without vicious circularity. One cannot appeal to the implicit beliefs or commitments that are constitutive of the operative faculties and subfaculties if these beliefs or commitments also comprise the worldview. As Sosa points out, however, this circularity is vicious only if the explicit beliefs or commitments arrived at through an epistemic perspective or worldview are identical with the implicit beliefs or commitments from which they are deduced as conclusions. The conclusions we reach are conscious, explicit beliefs that are constitutive of an epistemic perspective or worldview. For Descartes and Reid, Sosa argues that these conclusions are not the same, epistemically, as the premises or implicit beliefs or commitments that guide our rational conduct. Regardless of how closely related our premises and conclusions may be, and even if they turn out to share the same propositional contents, there is still a difference in the degree or quality of knowledge.

According to the PC principle, knowledge is enhanced through an epistemic perspective or worldview by justified trust in the reliability of its sources. In the transmutation of mere true belief into a higher and more desirable form of knowledge, Sosa [3, pp. 77–78] argues that something is added to the body of beliefs that comprise our epistemic perspective or worldview, whereby they are made more comprehensively coherent. Thus, Reid's principles of Common Sense, comprised of the implicit beliefs or commitments that guide our rational conduct, are justified because they lead us in an epistemically appropriate way rather than haphazardly to an epistemic perspective or worldview whose beliefs gain epistemic value by being part of a more comprehensively coherent framework. We judge our beliefs better justified through awareness of our cognitive constitution. From the vantage point of a conscious epistemic perspective, Sosa [3, pp. 79–80] argues that we can evaluate the success or reliability of our various faculties and subfaculties and the implicit beliefs or commitments they form in leading us aright. This implies that the principles of Common Sense or our presuppositions, implicit beliefs or commitments are not justified beforehand according to the legitimating account of justification stated by the KR requirement, but retrospectively from the vantage point of an epistemic perspective or worldview.

According to Sosa [3, p. 75], then, circularity at the level of reflective knowledge is plausibly prohibited because reflective knowledge goes beyond animal knowledge by integrating it in a more comprehensively coherent perspective. Through an epistemic perspective, object-level animal beliefs may be seen as reliably based not on reasons but on ground-level competences, and thereby transmuted into reflective knowledge. Here animal knowledge involves justification of a sort, namely, a first-order justification through a competence that is part of our natural endowment as rational beings. The sort of justification that has epistemic value, however, requires awareness of one's cognitive doings as in internalism and whether they have led us in an epistemically appropriate way. When a belief adds to the coherence of one's picture of the world in an interesting way, Sosa [3, p. 131] argues that one is able to gain an additional measure of epistemically valuable justification for one's beliefs. This additional measure of justification goes beyond the reliability of how we must acquire contents and form beliefs at the level of animal knowledge. It does so by bringing to consciousness a well-founded account of how our first nature and emplacement in the world yield such reliability. This internalist account of our cognitive doings, however, is undertaken retrospectively from the vantage point of an epistemic perspective rather than in advance of developing such a perspective.

The transmutation of mere true belief at the level of animal knowledge into reflective knowledge through an epistemic perspective requires that implicit beliefs or commitments be epistemically evaluable in at least some of the same ways as explicit beliefs or commitments. This is what gives beliefs, whether implicit or explicit, normative status. Like explicit beliefs or commitments, implicit beliefs or commitments can be well formed or not, and based on reasons despite being involuntary. Thus, Sosa [3, p. 233] rightly argues that there is no reason to suppose that such presuppositions should not share at least some of the same normative features of explicit beliefs or commitments. Accordingly, Sosa [3, p. 243] concludes that we should be able to use our faculties to gain access to knowledge of the ways in which and the extent to which these faculties are reliable, at least in rough outline.

According to virtue perspectivism, dispositions, faculties, or competences that are not reason-based are involved in animal knowledge when we acquire and sustain beliefs correctly. The deliverances of these animal competences, comprised of presuppositions, implicit beliefs or commitments, are constitutive of reason-based competences such as intellectual virtues. If we are to avoid vicious circularity, however, Sosa [3, p. 234] argues that animal competences need their own epistemically normative status. This normative status cannot be earned through reasoning based on the justified deliverances of these competences beforehand. The normative status of an explicit belief or commitment cannot derive from the presence and operation of an implicit belief or commitment, while the implicit belief or commitment acquires its normative status only after the explicit belief or commitment has its proper epistemic status.

In contrast to reason-based competences that form beliefs whose epistemic status is earned largely through one's own thought and effort, animal competences form beliefs whose epistemic status is earned largely through the brute reliability of these competences which function automatically in the absence of thought and effort. In reflective knowledge, the minimal epistemic status of animal beliefs can be boosted through an endorsing perspective on the reliability of the deliverances of animal competences and their sources. From the vantage point of an epistemic perspective or worldview, an implicit belief or commitment can be made explicit and defended against skeptical doubts concerning both the source or basis (competence) from which it is derived and the conditions or circumstances in which it is exercised. We can shape our natural cognitive practices, moreover, by enhancing their epistemic virtue. We can do so at least to some extent, Sosa [3, p. 142] argues, since an epistemic perspective does not require that every belief in our cognitive structure be freely chosen and deliberate.

References

1. Foley R (2012) The foundational role of epistemology in a general theory of rationality. In: Fairweather A, Zagzebski L (eds) Epistemic authority: essays on epistemic virtue and responsibility. Oxford University Press, New York, pp 214–230
2. Sosa E (2007) Apt belief and reflective knowledge. Vol. 1: A virtue epistemology. Oxford University Press, New York
3. Sosa E (2009) Apt belief and reflective knowledge. Vol. 2: Reflective knowledge. Oxford University Press, New York
4. Sosa E (1986) The raft and the pyramid: coherence versus foundations in the theory of knowledge. In: Moser PK (ed) Empirical knowledge: readings in contemporary epistemology. Rowman and Littlefield, Totowa, pp 145–170
5. Zagzebski L, Fairweather A (2001) Introduction. In: Fairweather A, Zagzebski L (eds) Virtue epistemology: essays on epistemic virtue and responsibility. Oxford University Press, Oxford, pp 3–14
6. Cohen S (2002) Basic knowledge and the problem of easy knowledge. Philos Phenomenol Res 65(2):309–329. doi:10.1111/j.1933-1592.2002.tb00204.x

Chapter 4
Challenges to Virtue Perspectivism

Sosa's conception of the relation between reason-based competences and animal competences that are not reason-based in arriving at an epistemic perspective suggests a more nuanced understanding of the relation between trust and rationality. Sosa argues that the right model for understanding reflective rational justification is not the linear model, where justification is transmitted from premises to conclusions along a unidirectional pipeline of reasoning. A better model for understanding the relation between animal competences and reason-based competences, as well as the corresponding relation between unreflective rational justification and reflective rational justification, is the web of belief. According to this model, the whole web or network of belief is properly attached to the environment through the causal mechanisms of perception and memory. Each belief in the web, or node in the network, gains its epistemic status partly through its mutual basing relations with other beliefs in the web or nodes in the network.

Sosa argues that this model helps us see that the reflective endorsement of our largely unconscious cognitive doings is not viciously circular. Through the bidirectional relation between one's commonsense animal knowledge and reflective scientific knowledge, one sees his modes of rational basing and other belief acquisition as sufficiently reliable and truth-conducive. The reflective endorsement of one's beliefs or commitments, which are implicit and unreflective in the first instance, lends an additional measure of epistemic justification to them. When we modify a belief or commitment, whether explicit or implicit, in the light of evidence from psychology, neurobiology, cognitive science or commonsense, we do so based on beliefs that are acquired through commitments already in place, most prominently through those commitments involved in perception. Of course, Sosa acknowledges that this gives rise to an ineliminable circle in the way we come to hold and modify our implicit and explicit perceptual commitments. We hold them and sustain them over time based on continuing observations, monitoring, and particular perceptual beliefs which are themselves based on our current and perhaps modified commitments. We now consider two challenges to this view of knowledge, one arising from internalism and the other arising from the lack of a

© The Author(s) 2014
M.G. Harvey, *Wireless Next Generation Networks*,
SpringerBriefs in Electrical and Computer Engineering,
DOI 10.1007/978-3-319-11903-8_4

47

clear distinction between animal knowledge and reflective knowledge in Sosa's 2-level basic knowledge structure.

4.1 Legitimation and Retrospective Justification

In response to the problem of epistemic circularity, Sosa [1, pp. 239–240] argues that there is nothing especially vicious about the beliefs in our web or the nodes in our network constituted by these commitments. If justification is undertaken retrospectively from the vantage point of an epistemic perspective on the reliability of our animal dispositions rather than in advance of such a perspective, we need not meet the following requirement of internalism for a legitimating account of a general theory of knowledge.

MR   In order to understand one's knowledge satisfactorily, one must see oneself as having some reason to accept a theory that one can recognize would explain one's knowledge if it were true.

This requirement states that one needs good reasons up front for accepting a theory of knowledge or, for that matter, for accepting a theory of rationality based on a theory of knowledge.

The MR requirement for a legitimating account of human knowledge is analogous to the KR requirement for a legitimating account of justification. According to the KR requirement, one needs to establish the reliability of a cognitive faculty such as perception before one is justified in trusting its deliverances. Similarly, according to the MR requirement, one needs to establish the reliability of a theory of knowledge in its account of our cognitive faculties before one is justified in trusting this account as a satisfactory explanation of how our knowledge is acquired and made more comprehensively coherent. According to internalism, moreover, the reasons adduced in support of our acceptance of this theory cannot be generated by the theory itself without vicious circularity.

Since internalism precludes *a priori* any sort of circularity as a possible way in which such reasons might be generated, Sosa [1, p. 196] argues that skepticism is a forgone conclusion. This can be seen from the following internalist argument.

P   There is no way to adequately support the belief that our faculties, which include perception, intuition, memory, deduction, abduction and testimony, are reliable without employing those faculties.

C   Therefore, there is no way to arrive at an acceptable theory of knowledge and its general sources.

Sosa rightly argues that the MR requirement assumes that the only valid account of knowledge is a legitimating account that shows how all human knowledge can be traced back to some epistemically prior knowledge, and how our current knowledge is inferentially derivable from this prior knowledge. According to Sosa [1, p. 172], the best argument against this assumption is that it has thus far not led to an account of knowledge that would satisfy its own requirement. That internalism cannot satisfy the MR requirement suggests that the requirement is incorrect or at least limited in its applicability.

Sosa's argument can be illuminated in the light of Kuhn's [2] conception of a normal paradigm in science. Like a normal paradigm which guides rational inquiry in science, the KR and MR requirements of internalism play a normative role in guiding rational inquiry in epistemology. Like scientists, epistemologists continue to adhere to the normal paradigm until the paradigm cannot solve all its problems. Even though the paradigm has led them into crisis, neither the scientist nor the epistemologist renounces their faith in the paradigm. According to Kuhn, this is what gives a normal paradigm its normative status and makes normal puzzle solving possible. Over time, however, accumulating doubts, anomalies, and evidence against the normal paradigm precipitates an epistemological crisis that opens up the possibility of extraordinary science which may lead to a paradigm shift. Crisis generates partial and fragmentary solutions to problems which the old paradigm could not solve. Sosa's virtue perspectivism can be seen as an attempt to combine such partial and fragmentary solutions into a more general account of human knowledge. Unlike many solutions in epistemology which tend to be narrowly focused, Sosa's solution aims to advance discussions in contemporary epistemology beyond current impasses by taking into account the partial and fragmentary solutions proposed by other leading theories of knowledge, and by integrating their insights in a more comprehensively coherent account of human knowledge. Thus, Sosa's general theory of knowledge is itself an instance of the theory he advocates.

The difference between a retrospective account of justification and a legitimating account is analogous to a paradigm shift in science. To help orient us in the right general direction, Sosa [1, p. 242] argues only that it must be acknowledged that the mutually supportive basing relations between our implicit and explicit beliefs or commitments on one hand, combined with the comprehensive coherence achieved through their interaction on the other, might add something of epistemic value. In this regard, Sosa's theory of knowledge shares several interesting features with Rescher's [3] theory of knowledge, most importantly Rescher's argument that justification is best undertaken retrospectively from an epistemic perspective on one's cognitive doings rather than in advance of such a perspective as required by internalism.

We need not meet the internalist requirement, either for a legitimating account of our animal competences before we are justified in trusting their deliverances, or for a legitimating account of human knowledge before we are justified in trusting the account as a satisfactory explanation of how our knowledge is acquired and made more comprehensively coherent. Instead, we need to see that these sources, bases, or competences already have a minimal sort of justification as they stand, namely, animal knowledge as apt belief. We need to see, moreover, that we can enhance the epistemic quality of their default justification retrospectively through having an epistemic perspective on the reliability of these animal competences and their deliverances. Such an epistemic perspective also allows us to monitor and revise implicit beliefs or commitments in cases where the relevant animal competences or dispositions did not reliably form beliefs that are truth-conducive.

4.2 The Relation Between Belief and Action

Perhaps a more serious challenge to Sosa's virtue perspectivism has to do with its neglect of the relation between belief and action. Like all virtue theories of knowledge, virtue perspectivism is motivated by an analogy between virtue epistemology and virtue ethics. Whereas virtue epistemology is focused on the cognizer and his beliefs, virtue ethics is focused on the agent and her actions. When the cognizer's beliefs qualify as knowledge or the agent's actions are right, Sosa [1, pp. 188–189] argues that they are so because of intellectual virtues or practical virtues, respectively, seated in the subject. These virtues give rise to the belief or action and make the cognizer or agent reliable and trustworthy over an interesting range of possible beliefs or choices. When we praise a cognitive performance as apt or an action as right, we assess not only the belief or action but the aptitude, skill, and character of the person who forms the belief or performs the action. Despite the emphasis on the subject, however, critics of virtue epistemology argue that virtue epistemologists are not so much interested in the intrinsic value of epistemic virtues as they are in how the conception of an epistemic virtue may have instrumental value in aiding their analysis of the properties of a belief such as justification. Wolterstorff [4], for example, argues that despite the emphasis on the properties of persons, virtue epistemology remains focused on the properties of beliefs. One might inquire, therefore, how the analogy between virtue epistemology and virtue ethics may be interpreted more substantively, where belief and action are seen as essentially related through the more general conception of virtue.

According to the analytic model of perception which informs contemporary epistemology, beliefs are mental states rather than acts. This claim is at the heart of internalism. The state of believing some proposition p in the case of the animal competence of perception, moreover, is not brought about by deciding to believe that p. One reacts to a visual experience as of a red wall, for example, by automatically forming the implicit belief or commitment that one sees something red. The subject is said to entertain the propositional content of the belief that p by holding one of two possible propositional attitudes toward p. One may ascent to p, or one may withhold ascent to p. Sosa argues that ascent, as well as withholding ascent, involves a second-order competence at the level of reflective knowledge. At the level of animal knowledge, however, implicit beliefs or commitments involve a sort of default ascent where this choice is lacking. Thus, Wolterstorff rightly argues that the analytic model of perception poses a formidable challenge for anyone who wants to let virtue ethics guide their thinking in epistemology. In response to this criticism, Sosa [5, p. 60] argues that such propositional attitudes that are not reason-based would still qualify as foundationally justified if justified by how they derive from a competence. Accordingly, Sosa [5, p. 51] concludes that virtue foundational justification is foundational justification that derives from the justified propositional attitude's manifesting an epistemic competence.

It is uncertain, however, whether this response can be applied generically to all perceptual experience. There are cases which do not fit the analytic model of perception well or at all. Consider the case, for example, where one is presented with a visual experience as of a red wall, but does not automatically form an implicit belief or commitment that one sees something red. Suppose that a structural engineer has marked a wall red on a construction site because it is unstable and thus unsafe for workers. In such circumstances, one might express one's reaction to the visual experience as of a red wall, not by forming an implicit belief or commitment that one sees something red, but by avoiding the wall because it is dangerous. In this case, one's reaction to the visual experience is expressed directly by one's behavior rather than one's belief. In different circumstances, however, where the lighting conditions are poor and one must decide whether a particular wall is marked red or not, one may form a belief based on his or her intellectual seeming. In this case, one's reaction to the visual experience is expressed indirectly by one's belief which seems necessary in order to judge whether the wall is red or not by holding the appropriate propositional attitude toward it. Here the belief not only plays a mediating role in one's visual experience, it involves a refinement of the visual experience itself at a more reflective level which is not required in the former case and may even pose a danger by delaying one's instinctive reaction to avoid the unsafe wall.

What is troubling about the analytic model of perception is the mediating role and inferential character of implicit beliefs or commitments, whether their source is an epistemic competence or not. Kuhn [2, p. 195] argues that since Descartes modern philosophy has analyzed perception—the animal mechanism responsible for the formation of implicit beliefs or commitments—as an interpretative process, as an unconscious version of what we do after we have perceived as in the case of explicit beliefs or commitments at the level of reflective knowledge. Although the analytic model of perception applies to vision, which depends on the animal competence of perception but also involves higher-order, reason-based information-processing competences, it is doubtful whether the model can be applied to the animal competence of perception itself. Sosa acknowledges this problem when he argues that the propositional content of implicit beliefs or commitments does not appear to involve general propositions of the form, "whenever I seem to see something red, I do see something red," or "if one has the visual experience as of seeing something red, then one sees something red." The analytic model of perception, which describes more or less correctly the conscious formation of explicit beliefs or commitments at the level of reflective knowledge, misrepresents the unreflective status and non-inferential character of implicit beliefs or commitments formed unconsciously at the level of animal knowledge. Thus, Sosa suggests that one's reaction or implicit belief or commitment to the visual experience as of seeing something red is perhaps better expressed by a "tendency" or "habitual" proposition. Such a proposition, Sosa [1, pp. 240–241] tells us, may be interpreted as a reliability claim of the form, "reliably, if I have a visual experience as of seeing something red, I tend to see something red."

On Sosa's account of animal knowledge, however, the awareness of risk or vulnerability concerning the reliability of one's implicit beliefs or commitments is precluded by the unreflective status and non-inferential character of animal perception. Although a "tendency" proposition makes sense within the context of reflective knowledge, it becomes problematic in the context of animal or unreflective knowledge. In particular, Sosa's suggestion of a "tendency" proposition in the case of perception at the level of animal knowledge introduces an ambiguity as to whether animal knowledge is completely unreflective or only partially so, and to what degree or extent. In order to avoid both the infinite regress and vicious circularity, as we have seen, one's beliefs must be derived in the first instance from sources or competences that are not reason-based but nevertheless rational with their own distinctive type of epistemic justification because they lead us aright. But, on occasion, Sosa suggests the opposite. For example, in the context of discussing animal knowledge, Sosa [1, p. 215] argues that a necessary condition for the formation of a belief is that one must *always* understand the proposition believed.

Although understanding is an intellectual virtue that is necessary for reflective knowledge, it poses a problem for the unreflective status of an animal disposition or competence. As we have also seen, Sosa argues that skepticism is best understood as an attack against reflective knowledge rather than animal knowledge. Thus, we may stop the pure regress by basing our belief-forming and belief-sustaining dispositions on animal competences that are not reason-based, and we may avoid the pure circle by showing how reflective knowledge requires an epistemic perspective or worldview in the light of which our animal competences may be seen as reliable and truth-conducive. But if animal knowledge involves understanding, and hence reflective knowledge at least to some degree, it is difficult to see how we can stop the pure regress or avoid the pure circle.

We need to connect belief and action in a circle that is virtuous rather than vicious. As we have seen, Sosa [1, p. 232] defines the implicit beliefs or commitments formed through our animal competences as implicit mental states of a subject S, who takes for granted or presupposes that his situation is propitious and his source generically reliable. These states are contingent in that they are ones that S can be in or not be in, and they have propositional content which makes them epistemically evaluable. This analytic definition of implicit belief or commitment assumes that belief is more basic or primary than action. Recall that, according to internalism, beliefs are mental states rather than acts. Unlike acts, beliefs are capable of transmitting justification from premises to conclusions and conversely in the case of mutual basing relations. But, as Wolterstorff [4] has pointed out, the emphasis on the evaluation of beliefs in contemporary epistemology leads to an inevitable disconnect with virtue ethics with its emphasis on the evaluation of actions.

What is needed is a definition of implicit belief or commitment that is more consistent not only with virtue ethics but with externalism. Such a definition would involve an inversion of belief and action, so that it is not belief and our mental states that are basic or foundational to rationality as in internalism but our reactions and attitudes toward the things with which we interact as in externalism, including other people and our environment. There is no reason to suppose that our implicit commitments must be restricted to and expressed by beliefs. It might also be the case that there are ground-level reactions underlying our beliefs that motivate and influence them, but which are not themselves beliefs. Such an inversion, which emphasizes the primacy of action over belief, would make virtue epistemology (belief) depend on virtue ethics (action) in a more substantive way as suggested by their analogy. This inversion would also allow us to see how our beliefs are extensions and refinements at a more reflective level of basic or ground-level reactions connected to our instinctual rational conduct at the level of animal knowledge. It would allow us to see how our conscious intellectual conduct is a refinement of our unconscious instinctual conduct, both of which are integral components of human rationality.

To understand the nature of this inversion, we might reconsider Wittgenstein's model of our cognitive structure which Sosa argues is an example of cognitive quietism. Sosa bases his conclusion on Strawson's [6] interpretation of Wittgenstein's model which may not accurately reflect Wittgenstein's own position. A key passage in *On Certainty* suggests the opposite conclusion. Toward the end of his life, Wittgenstein [7, Sects. 96–97] wrote that "it might be imagined that some propositions, of the form of empirical propositions, were hardened and functioned as channels for such empirical propositions as were not hardened but fluid; and that this relation altered with time, in that fluid propositions hardened, and hard ones became fluid. The mythology may change back into a state of flux, the river-bed of thoughts may shift. But I distinguish between the movement of the waters on the river-bed and the shift of the bed itself; though there is not a sharp division of the one from the other."

This metaphor or model of our cognitive structure corresponds to Sosa's distinction between two classes of beliefs or commitments, one implicit and the other explicit, both of which are constitutive of an epistemic perspective or worldview. Whereas the river-bed of thoughts corresponds to the implicit beliefs or commitments automatically formed through animal competences that are not reason-based, the movement of the waters (thoughts) on the river-bed corresponds to the explicit beliefs or commitments formed through reason-based competences such as intellectual virtues. This distinction corresponds in turn to Sosa's distinction between animal or unreflective knowledge and reflective knowledge. The river-bed of thoughts, or the hardened implicit beliefs or commitments that lie at the foundation of one's present epistemic perspective, function as channels or control beliefs that guide our rational conduct at the level of reflective knowledge as in Reid's normative principles of Common Sense.

There are at least three points worth considering in Wittgenstein's model of our cognitive structure. First, there is a movement both of the waters (thoughts) on the river-bed and a shift of the river-bed of thoughts itself as the beliefs or commitments of one class interchange their epistemic status and rational justification with the beliefs and commitments of the other class. The change in epistemic status has to do with which type of knowledge is involved, animal knowledge or reflective knowledge. The change in rational justification has to do with which type of justification is involved, unreflective rational justification or reflective rational justification. That the relation between these two classes of beliefs or commitments may change over time, accompanied by a corresponding change in their epistemic status and rational justification, suggests that the implicit beliefs or commitments that we presently hold, presuppose, or take for granted and find irresistible and necessary can change. Some propositions, of the form of empirical propositions, that were once hardened and functioned as channels to guide our intellectual conduct, may become fluid as they rise from the lower level of animal knowledge to the higher level of reflective knowledge where they are subject to epistemic evaluation, conscious monitoring, and control. Thus, like Sosa's virtue perspectivism, one's implicit beliefs or commitments can be made explicit through an epistemic perspective. If this parallel is right, we may question whether Wittgenstein's model of our cognitive structure is an example of cognitive quietism as Sosa claims. For cognitive quietism implies that implicit beliefs or commitments remain immune from criticism at a reflective level, and that one can never become aware of them in a way that could lead to their acceptance, rejection, or revision as required by internalism.

Secondly, since the relation between these two classes of beliefs or commitments may change over time in that fluid propositions might become hardened and hard ones might become fluid, the mutual basing relations can change. Similarly, Sosa argues that an epistemic perspective on one's cognitive doings requires not only logical coherence among our beliefs at the level of reflective knowledge but mutual basing relations. Cross-level coherence from the object at the level of animal knowledge to one's meta-belief about the object at the level of reflective knowledge, and conversely, is a special case of mutual basing relations. Coherence has both a horizontal dimension which involves the relation between one's meta-beliefs at the level of reflective knowledge, and a vertical dimension which involves the relation between one's implicit and explicit beliefs or commitments at the levels of animal knowledge and reflective knowledge respectively. Through cross-level coherence, Sosa [1, p. 243] argues that animal belief is guided by meta-belief, whereby we base the former on the latter. This mutual basing relation, however, may change. Whereas an animal belief may serve as a basis or source for its relevant meta-belief at one time, the meta-belief may serve as a basis or source for the evaluation of that animal belief or another animal belief at another time which may result in the acceptance, rejection, or revision of the animal belief.

Thirdly, there is no sharp division between the river-bed of thoughts and the waters (thoughts) moving on the river-bed. There is no clear distinction between one's implicit and explicit beliefs or commitments. Thus, it is difficult to draw a line above which lie the degrees of comprehensive coherence that are sufficient for knowledge. Similarly, Sosa argues that it is difficult to determine a threshold for an epistemic perspective. Through animal-level cognitive processing, Sosa argues that a comprehensively coherent perspective can come together which can underwrite the

continued use of those faculties. Accordingly, Sosa [1, p. 150] argues that gener-
alizations about the reliability of one's faculties arise from the combined operation of
past perception and memory and perhaps a gradual induction over time. This account
of the transition from animal knowledge to reflective knowledge, however, leads to a
problem familiar to Hegel scholars. It is not clear how the qualitative change from
animal (immediate) knowledge to reflective (mediate) knowledge is achieved.
Analogous to Hegel's [8] solution, Sosa suggests that the qualitative difference,
epistemically, can be bridged quantitatively by thinking of the quality of one's
knowledge as a matter of degree. That there can be no sharp division between
implicit and explicit beliefs or commitments leads to an inevitable blurring of the
distinction between animal or unreflective knowledge and reflective knowledge.

Like Wittgenstein, Sosa [1, p. 232] acknowledges this problem when he argues
that we need a distinction between the aptness of a belief at the level of animal
knowledge and the aptness of a belief at the level of reflective knowledge. Although
beliefs at both levels are skillfully acquired through competences, animal competences such as basic trust operate at an unconscious level in the dark, while reason-
based competences such as intellectual virtues operate at a conscious level in the
light of our awareness of animal beliefs and control over them. Despite this rec-
ognition, and the fact that most of the time Sosa speaks of animal-level cognitive
processing as an unconscious and unreflective process, there are times such as
above when he suggests the opposite. Here, the impossibility of a sharp division
between animal or unreflective knowledge and reflective knowledge, combined
with the implication that some degree of reflective knowledge might be involved in
animal knowledge, invites vicious circularity.

The parallels drawn above between Wittgenstein's model of our cognitive
structure and Sosa's virtue perspectivism might provide a way beyond the impasse.
Wittgenstein fleshes out his model of our cognitive structure in several places. In
one place, Wittgenstein [7, Sect. 330] argues that the statement "I know that p"
expresses the readiness to believe certain things. Similarly, Reid calls the readiness
to believe a credulity disposition. Wittgenstein [7, Sect. 235] argues that a credulity
disposition, however, may misrepresent the epistemic status of this disposition:
"that something stands fast for me is not grounded in my stupidity or credulity."
A credulity disposition implies that one might have acquired one's animal beliefs
or implicit beliefs or commitments haphazardly, in contrast to an epistemically
appropriate way. For both Wittgenstein [9, Part I, Sect. 219] and Reid [10, p. 9],
it is not within our power or control to choose which beliefs we will accept, reject,
or revise at the level of animal knowledge. Rather, we are attracted to them as if
they were irresistible and necessary.

Thus, Wittgenstein [7, Sect. 475] argues that the human being should be regarded
in the first instance as an animal, as a primitive being to which one grants instinct but
not reason. Like lower animals, human beings have a natural instinct or reaction to
trust their surroundings and other members of their group. Through experience and
interaction with one's environment and other people, however, one may learn to
distrust as the level of risk or vulnerability increases. Wittgenstein [7, Sects. 115, 354]
suggests that doubting and reasoning are refinements of the more primitive language

game of trusting, both of which are necessary for us to survive and to arrive at a more accurate and comprehensively coherent perspective, framework, or worldview. Whereas trust is a natural and spontaneous reaction to believe that *p* at the level of animal knowledge, Wittgenstein [7, Sect. 283] and Reid [10, pp. 281–282] suggest that doubt is something we must learn to do at the level of reflective knowledge. Wittgenstein [7, Sect. 322] asks, for example, "what if the pupil refused to believe that this mountain had been there beyond human memory? We should say that he had no *grounds* for this suspicion." In terms of Sosa's virtue perspectivism, we should say that he had no basis or source for justifiedly believing that such implicit beliefs or commitments were unreliable and might lead him astray in his cognitive doings.

According to Wittgenstein [7, Sect. 144], we acquire a system of implicit beliefs or commitments through the natural reaction of trust, which functions as a foundation to guide our reasoning at a more reflective level in arriving at an epistemic perspective or worldview. Thus, Wittgenstein argues that the child learns to believe a host of things through this natural mechanism or competence. Bit by bit, a system of what is believed is formed, namely, our cognitive structure. In this system, some beliefs or commitments stand unshakably fast as in the river-bed of thoughts. Some propositions, of the form of empirical propositions, have become hardened and function as channels for such propositions as are not hardened but fluid. Whereas these implicit beliefs or commitments guide our intellectual conduct, other beliefs or commitments in our cognitive structure are more or less liable to shift as in the waters (thoughts) moving on the river-bed. These propositions, also of the form of empirical propositions, consist of the explicit beliefs or commitments that we arrive at through having an epistemic perspective on our belief-forming and belief-sustaining mechanisms. In thus acquiring and forming his beliefs through a trustful disposition, Wittgenstein [7, Sect. 538] argues that the child learns to react in a certain way, and in so reacting it does not so far know anything. Knowing, in the sense of reflective knowledge, begins only at a later stage. Elsewhere, Wittgenstein [11, p. 31] argues that "the origin and the primitive form of the language game is a reaction; only from this can more complicated forms develop. Language—I want to say—is a refinement, 'in the beginning was the deed.'"

Wittgenstein's [7, Sect. 131] model of our cognitive structure suggests that one's epistemic perspective or worldview is based on natural reactions or shared agreements in judgment, in contrast to shared agreements in opinion. Unlike shared agreements in opinion, which involve explicit beliefs or commitments formed through reason-based competences such as intellectual virtues, natural reactions or shared agreements in judgment involve implicit beliefs or commitments formed through animal dispositions or competences that are not reason-based. According to Wittgenstein [11, p. 46; 9, Part I, Sect. 546], both our implicit and explicit beliefs or commitments are extensions and expressions of ungrounded natural reactions. Wittgenstein [7, Sects. 358–359] describes this distinctive type of animal disposition or competence with its characteristic certainty as a form of life: "I would like to regard this certainty, not as something akin to hastiness or superficiality, but as a form of life ... that means that I want to conceive it as something that lies beyond being justified or unjustified; as it were, as something animal." For Wittgenstein, certainty is not an outcome of reflective knowledge, the quality of one's knowledge, or an inference that can be drawn from the explicit beliefs or commitments one holds. Instead, certainty is a property of the natural reactions themselves that form our implicit beliefs or commitments. These beliefs or commitments are certain because they arise from a reliable and irresistible cognitive mechanism or competence.

If this account of Wittgenstein is right, we might question the validity of Sosa's argument that certainty, like knowledge, is a matter of degree which can be enhanced through having an epistemic perspective. According to Sosa [1, p. 143], just as we know some things better or more justifiedly than other things through having an epistemic perspective on those things, so we are more certain of some things than other things because we know them in this more epistemically valuable way. Presumably, cross-level coherence has the epistemic power to impart certainty in addition to justification by appealing to its reliability as a source of what renders it certain. Thus, Sosa argues that we know things with a better assurance in the light of reflective knowledge. In contrast, Wittgenstein argues that certainty is a property of the natural reactions that form one's implicit beliefs or commitments. This certainty, moreover, is lacking in the reflective knowledge that forms our explicit beliefs or commitments. According to Wittgenstein [7, Sect. 308], reflective knowledge and animal certainty belong to different categories. They are not two mental states such as surmising and being sure. Instead, certainty is part of an ungrounded reaction or animal competence that lies at the foundation of reflective knowledge and its necessary companion, doubt. Knowing that p in the sense of unreflective knowledge is a natural reaction unaccompanied by doubt, which consists in a way of acting according to a system of implicit beliefs or commitments with assurance. For this reason, we find the deliverances of this animal disposition irresistible and necessary.

Knowing that p in the sense of reflective knowledge, however, depends on a second-order competence, whereby we are able to ascent to p or withhold ascent to p. This second-order competence requires doubt in order to judge when it is appropriate to give ascent and when it is appropriate to withhold ascent when there is too much risk of error or vulnerability, epistemically. Thus, according to Wittgenstein [7, Sects. 166, 204], reflective knowledge and doubt are deeper ways of believing, both of which may be seen as extensions and refinements of our natural reaction to trust the reliability of our belief-forming and belief-sustaining dispositions or competences. The difficulty is to realize the groundlessness of natural reactions. Here, as Wittgenstein [7, Sect. 471] argues, "it is so difficult to find the *beginning*. Or, better: it is difficult to begin at the beginning. And not try to go further back." When we try to justify the animal disposition or competence of trust itself, the type of rational justification available at the level of reflective knowledge through having an epistemic perspective is always less certain than the first-order, unreflective rational justification available at the level of animal knowledge.

This implies that there are limits to reflective rational justification. Although we can justify the deliverances of the trust disposition at the level of reflective knowledge through an epistemic perspective on the reliability of the disposition, we cannot justify the disposition itself. Although an epistemic perspective can justify our trust in the reliability of other animal dispositions or faculties that depend on the animal disposition of trust such as perception and memory, it cannot justify the basic trust disposition itself. Thus, we need to redefine the PC principle in the light of this limitation.

PC*   Knowledge is enhanced through justified trust in the reliability of its sources, which depend on natural reactions that lie beyond being justified or unjustified.

This new version of the principle makes it clear that it is not the basic trust disposition itself that needs justification, but rather the implicit beliefs or commitments formed by it. An epistemic perspective is capable of transmuting what is believed through the disposition into knowledge, not the disposition itself. In the transmutation of what is believed through a trustful disposition into a higher and more desirable state of knowledge, trust itself remains foundational to rationality and is unaffected by any such epistemic enhancement. Although the implicit beliefs or commitments formed through animal competences rooted in natural reactions are capable of transmitting justification from premises to conclusions, the natural reactions themselves need not have this capability since they are actions rather than beliefs. For this reason, they are neither justified nor unjustified. Instead, Wittgenstein argues that they are constitutive of our first nature which he calls a form of life.

4.3 A 3-Level Basic Knowledge Structure

We can enhance Sosa's 2-level basic knowledge structure with a 3-level basic knowledge structure that allows us to connect belief and action in a more substantive way commensurate with a virtue approach to knowledge. Figure 4.1 shows how the primacy of belief is replaced by the primacy of action in the foundations of the knowledge structure. The 2-level triangle in the structure represents two classes of

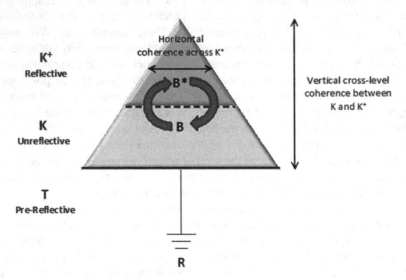

Fig. 4.1 A 3-level basic knowledge structure

beliefs, one implicit and the other explicit, hereafter denoted by B and B*
respectively. The boundary line separating these two classes of beliefs is represented
by a dashed line to indicate that there is no sharp division between them. In contrast,
the boundary line separating basic trust T from animal knowledge K is represented
by a solid line to indicate that there is a clear distinction between pre-reflective
epistemic states (reactions) and unreflective epistemic states (implicit beliefs) which
might contain some small element or degree of reflection K$^+$ as hinted by Sosa. The
triangle is wider at the bottom to indicate that most of our basic knowledge structure
is comprised of implicit beliefs or commitments that have not been transmuted into
knowledge through having an epistemic perspective on the reliability of our animal
beliefs and their sources at the level of reflective knowledge. Only a relatively small
fraction of animal beliefs have been thus transmuted at the top of the triangle.

Although beliefs may be based on other beliefs as sources for their justification
through mutual basing relations, all beliefs, whether reason-based or not, are rooted
in natural reactions R that are constitutive of our first nature. B* consists of meta-
beliefs or explicit beliefs or commitments formed through reason-based competences such as deduction, abduction, and intellectual virtues. This class of beliefs is
necessary for an epistemic perspective. In contrast, B consists of object-beliefs or
implicit beliefs or commitments formed through animal competences that are not
reason-based such as perception, memory, and various kinds of procedural knowledge that include inferential patterns encoded in our instinctual rational conduct.

According to this basic knowledge structure, while B* derives from B, B derives
its justification from B* in a circular fashion. B, moreover, derives from R as shown
in the 3-level tree of the structure. The implicit beliefs or commitments formed
through animal competences derive in turn from a distinctive type of animal
competence T, namely, basic trust in the reliability of the other ground-level animal
competences such as perception and their deliverances. T belongs to a special class
of ground-level animal competences R, comprised of natural reactions or shared
agreements in judgment that are constitutive of our first nature. For this reason,
Wittgenstein calls these natural reactions forms of life. The class R has four
properties. First, natural reactions are neither justified nor unjustified, and thus
require no mutual basing relations so that they can be considered properly basic.
They are natural in that they are constituted by nature. Secondly, natural reactions
are inescapable, irresistible, and necessary. We cannot choose whether to act or not
act in accordance with them. Their function is to guide our rational conduct at an
instinctual level. Thirdly, natural reactions are characterized by certainty. We act
according to a system of belief formed through the animal competences that depend
on R with assurance until we have reason to suspect that a particular belief has not
been reliably formed or is not truth-conducive. At the level of animal knowledge,
trust precludes doubt since doubt depends on reflection which is plausibly prohibited in R by its pre-reflective status. Fourthly, natural reactions, though not
justified themselves, are nevertheless rational because to obsess over the sources or
bases of our knowledge is not rational as Descartes argued.

Two consequences follow from the properties of R. First, B derives or inherits its
characteristic certainty from the roots of the tree in R rather than from the leaves

of the tree in B*. Certainty comes from "below" in a pre-reflective fashion rather than from "above" as an outcome of reflection. Secondly, since both reason-based competences and competences that are not reason-based depend on T as well as the explicit and implicit beliefs or commitments formed by them, B* and B respectively, and since T requires no mutual basing relations, the infinite regress is stopped.

This still leaves the problem of vicious circularity. The 3-level basic knowledge structure advances Sosa's 2-level basic knowledge structure based on the distinction between animal knowledge and reflective knowledge by adding a distinction between belief and action at the level of animal knowledge. Whereas implicit animal commitments are beliefs, the shared agreements in judgment underlying and motivating them are natural reactions. Thus, our epistemic perspective or world-view at the level of reflective knowledge is "rooted" in some kind of action or reaction rather than "based" on some kind of belief at the animal level. The distinction between implicit beliefs or commitments B and natural reactions R, which involve two fundamentally different types of animal competences, allows us to make the following argument against the infinite regress and vicious circularity.

P1   K is justified, not by being based on T, but by being based on K^+ which requires having an epistemic perspective on K so that T need have no justification of its own.

P2   Skepticism attacks K^+, not K.

C   Therefore, by "rooting" the epistemic circle in T rather than "basing" it on T, where K^+ is plausibly prohibited in K by the pre-reflective status of T, we may avoid both the infinite regress and vicious circularity.

Whereas the first premise P1 defends against cognitive quietism, premises P1 and P2 jointly held defend against the pure regress and the pure circle. The key point of this 3-level basic knowledge structure is that natural reactions, unlike the implicit beliefs or commitments that are motivated by them, do not pose the problem of determining a threshold between animal or unreflective knowledge K and reflective knowledge K^+ which haunts Sosa's 2-level basic knowledge structure. There is a clear distinction between pre-reflective knowledge and reflective knowledge which allows the possibility that there might be some small element or degree of reflective knowledge in animal knowledge, but in a way that does not lead to vicious circularity.

Unlike behavior, which may also express our natural reactions unaccompanied by belief, beliefs are refinements of our natural reactions in a way that behavior is not. Unlike behavior, beliefs are capable of transmitting justification from premises to conclusions through horizontal and especially cross-level coherence. Perhaps this is why beliefs are considered more primary than actions in contemporary epistemology. Because beliefs allow us to guide our animal commitments, they have become the hallmark of rationality. Whereas B involves a refinement of R, which consists of implicit states that are evaluable at a higher epistemic level K^+, B* involves a further refinement of B, which consists of beliefs derived from B that have been epistemically enhanced or better justified through having an epistemic perspective on the reliability of B and its sources.

What must not be forgotten, however, is that both classes of beliefs, implicit as well as explicit, are rooted in reactions which are not themselves justified but are nevertheless foundational to rationality. In contrast to Sosa, this assumes that we can conceive of these reactions as sources of knowledge without claiming that they are also bases for knowledge. Thus, while our implicit beliefs or commitments derive their justification from explicit beliefs or commitments at a higher reflective level in a retrospective fashion, and this mutual basing relation may change over time as implicit and explicit beliefs interchange their epistemic roles within the circle, both classes of beliefs or the circle itself are rooted in instinctual reactions rather than based on more basic beliefs.

Far from leading to cognitive quietism, Wittgenstein's model of our cognitive structure allows us to draw an additional distinction between two types of animal competences which helps us avoid cognitive quietism, the infinite regress, and vicious circularity. The natural disposition or reaction to trust the reliability of animal competences such as perception makes reflective knowledge and an epistemic perspective possible. At the same time, the cultivation of intellectual virtues through reason-based competences at the level of reflective knowledge makes the justification of the reliability of our animal competences possible through an epistemic perspective and the cross-level coherence between our implicit and explicit beliefs or commitments.

Both types of competences, whether animal or reason-based, are rooted in basic trust which is itself a distinctive type of animal competence or disposition that requires no justification of its own. Thus, animal competences such as perception are not justified by our trust in them, but through having an epistemic perspective in the light of which we can see how the beliefs formed by trusting our animal competences are reliable and truth-conducive. Contrary to critics of virtue epistemology, Sosa's virtue perspectivism is not just a new move in an old game. Although the old elements are still there, they are integrated in a way that might lead to a new fertile point of departure for the old game in virtue ethics. But the foundation for the game needs to be changed from beliefs to actions to avoid vicious circularity and to make the approach more commensurate with a virtue approach to knowledge.

References

1. Sosa E (2009) Apt belief and reflective knowledge. Vol. 2: Reflective knowledge. Oxford University Press, New York
2. Kuhn TS (1970) The structure of scientific revolutions, 2nd edn. University of Chicago Press, Chicago
3. Rescher N (1992) A system of pragmatic idealism. Vol. 1: Human knowledge in idealistic perspective, Chaps 2, 4, 10. Princeton University Press, Princeton

4. Wolterstorff N (2003) Review of Virtue epistemology: essays on epistemic virtue and responsibility. Ethics 113(4):876–879
5. Sosa E (2007) Apt belief and reflective knowledge. Vol. 1: A virtue epistemology. Oxford University Press, New York
6. Strawson PF (1985) Skepticism, naturalism and transcendental arguments. In: Skepticism and naturalism: some varieties. Methuen, London, pp 1–25
7. Wittgenstein L (1969) Anscombe GEM, von Wright GH (eds) On certainty (trans: Paul D, Anscombe, GEM). Harper Torchbooks, New York
8. Hegel GWF (1975) Hegel's logic: being part one of the encyclopaedia of the philosophical sciences (trans: Wallace W), 3rd edn. Clarendon Press, Oxford
9. Wittgenstein L (1958) Philosophical investigations (trans: Anscombe, GEM), 3rd edn. Macmillan, New York
10. Reid T (1983) Beanblossom RE, Lehrer K (eds) Thomas Reid's inquiry and essays. Hackett, Indianapolis
11. Wittgenstein L (1980) Von Wright GH, Nyman H (eds) Culture and value (trans: Winch P). University of Chicago Press, Chicago

Chapter 5
Other Theories of Trust and Trust Models

We have discussed how animal competences such as perception depend on a distinctive type of animal competence called basic trust, and how reason-based competences such as intellectual virtues can be used to monitor, control, and moderate basic trust dispositions at a reflective level. We have discussed how our 3-level basic knowledge structure allows us to relate trust and rationality in a non-circular fashion by replacing the primacy of belief with the primacy of action in the foundations of the structure. We have also discussed how both types of competences, whether animal or reason-based, depend on natural reactions or shared agreements in judgment, not themselves justified but nevertheless foundational to rationality. That both types of competences depend on shared agreements in judgment means that knowledge is social from the start. Thus, we need not construct sophisticated arguments that show how we can extend trust in the reliability of our own faculties to trust in the reliability of the faculties of other people.

In this chapter, we discuss how individual competences can be generalized to social competences, and how individual cognitive structures such as our 3-level basic knowledge structure can be generalized to groups or systems. In particular, we discuss how our 3-level basic knowledge structure can be integrated with other theories of trust to yield a more enlightened perspective on the relation between trust and rationality. We begin with Zagzebski's theory of trust which draws two distinctions, one between self-trust and trust in other people and another between basic trust and full-fledged trust. Zagzebski employs these distinctions in developing the notion of rational trust or being properly trusting. We then return to Foley to discuss his definition of responsible belief which is based on the more restrictive and rigorous notion of epistemic rationality or normativity exemplified by Sosa's virtue perspectivism.

Foley's definition of responsible belief allows us to develop a non-circular theory of rationality by anchoring the definition in the notion of epistemic rationality or normativity. This definition of responsible belief also allows us to place practical limitations on the time and intellectual or computational resources required to pursue the intellectual goals of now having accurate and comprehensively coherent beliefs and full-fledged trust given a context. Thus, Foley's definition of

© The Author(s) 2014
M.G. Harvey, *Wireless Next Generation Networks*,
SpringerBriefs in Electrical and Computer Engineering,
DOI 10.1007/978-3-319-11903-8_5

responsible belief can be used to integrate Sosa's rigorous account of epistemic
normativity with Zagzebski's conception of full-fledged trust in a unified theory of
trust. This unified theory, with appropriate modifications from our 3-level basic
knowledge structure, can be translated into a practical, virtue-based trust model that
can be applied in a social context. This virtue-based trust model helps us understand
how to achieve socially valuable ends such as cooperation and collaboration based
on epistemically appropriate states through the exercise of intellectual virtues. At
the same time, it helps us avoid socially undesirable ends such as selfishness and
conflict based on epistemically inappropriate states.

5.1 Self-Trust and Trust in Other People

Zagzebski [1] helps us see how belief and action might be connected in a more
substantive way that is commensurate with a virtue approach to knowledge. Like a
moral virtue, Zagzebski argues that an intellectual virtue has both a motivational
component and a component of reliable success in attaining the end of the motivational component. Actions are performed by agents who exert powers and
typically bring about an effect or end through the exercise of those powers. A power
is not a power unless it is reliably effective in bringing about an end. In the light of
Aquinas' definition of virtue as the perfection of power, Zagzebski argues that the
agent need only exercise a power, not necessarily a virtue, that brings about the end
of the motivational component. Thus, an intellectual virtue qualifies as knowledge
if one exercises a power that attains its end because of this power which requires the
exercise of agency. Since Kant, agency has been understood as a necessary condition for autonomy, where autonomy has been defined as a necessary condition for
rationality. Thus, if trust and rationality are to be integrated in a unified theory of
trust, we need to show how trust is compatible with autonomy and agency.
According to internalism, only conscious agents have agency. The question is
whether a causal capacity or belief-forming disposition or competence such as basic
trust can be a power that requires the exercise of agency.

Zagzebski's response to this question is influenced partly by Korsgaard's [2]
interpretation of Kant's conception of the relation between reason and desire, and
more substantively by Sosa's [3, 4] virtue perspectivism. According to Korsgaard,
Kant suggests that autonomy is compatible with acting out of desire so long as the
reflective mind endorses the bidding of desire. Thus, we can be autonomous even
when we act instinctually. Similarly, Zagzebski argues that we can be autonomous
rational agents even in the simplest perceptual knowledge so long as we endorse the
bidding of our pre-reflective minds. This suggests that agency can be extended to
animal competences. According to Zagzebski, we need not assume that agency cannot
apply to an epistemic state that is initially acquired non-voluntarily. For if the state is
acquired in a non-voluntary way as in the case of implicit beliefs or commitments

formed by animal competences such as perception, the agent's subsequent reflection about her epistemic state at the level of reflective knowledge makes it voluntary.

Zagzebski constructs several arguments for this claim which involve parallels with Sosa's virtue perspectivism. If we reject the Aristotelian distinction between voluntary and non-voluntary actions, Zagzebski claims that the concept of agency can be extended to those parts of our cognitive structure that are initially acquired non-voluntarily. Accordingly, since animal dispositions or competences such as perception fall within this category, Zagzebski [1, pp. 153–155] argues that agency can be extended to the formation of our perceptual beliefs. In place of Aristotle's sharp distinction between voluntary and non-voluntary actions, Zagzebski argues that we may classify actions according to the degree or quality of conscious awareness and control on the part of the agent concerning her actions. This argument is analogous to Sosa's conception of knowledge and certainty as matters of degree that correspond to the quality of one's knowledge. Zagzebski [1, pp. 142–143] argues that one class of such actions consists of fully deliberate actions that are preceded by conscious deliberation and choice. These actions fall within a larger class of intentional actions that include actions that are not preceded by conscious deliberation and choice, and a still larger class of actions that are non-intentional but are nevertheless subject to moral evaluation. This argument is analogous to Sosa's conception of a class of implicit beliefs or commitments that are non-inferential but are nevertheless subject to epistemic evaluation. Thus, if we interpret the belief-forming and belief-sustaining disposition of basic trust as a non-voluntary part of our cognitive structure that is nevertheless subject to moral evaluation, we can relate trust and rationality through the central concept of agency.

Not surprisingly, Zagzebski's conception of the role of agency in implicit belief suffers from the same problem as Sosa's virtue perspectivism. There is a blurring of the distinction between animal knowledge and reflective knowledge. Analogous to the problem in Sosa's virtue perspectivism of determining a threshold for an endorsing perspective on our animal beliefs absent a clear distinction between animal or unreflective knowledge and reflective knowledge, there is a problem in Zagzebski's conception of the role of agency in implicit belief of determining a threshold for an endorsing perspective on our actions absent a clear distinction between non-voluntary and voluntary actions such as the one Aristotle makes. For Zagzebski, actions correspond to beliefs which reflect varying degrees of conscious awareness and control over the sources of our actions and their reliability. But if actions are smeared out over a continuum or gradient of conscious awareness and control on the part of the agent, it is difficult to see where animal or unreflective knowledge leaves off and reflective knowledge begins. Since skepticism attacks reflective knowledge rather than animal knowledge according to Sosa, the problem of vicious circularity appears once again.

Nevertheless, Zagzebski's [5, p. 17] theory of trust draws two distinctions that help us translate our 3-level basic knowledge structure into a practical, virtue-based trust model that can be applied in a social context. The first distinction concerns two types of basic trust, self-trust and trust in other people, hereafter denoted by T and T* respectively. Zagzebski defines self-trust as a disposition or attitude toward one's own cognitive faculties with three dimensions, including belief, feeling, and behavior. In the case of the cognitive dimension of trust, one believes that her faculties are generically trustworthy for attaining truth in the light of the awareness of being vulnerable to falsehoods. In the case of the affective dimension of trust, one has a feeling of trust toward one's own faculties in being able to achieve this intellectual goal. In the case of the behavioral dimension of trust, one treats their own faculties as trustworthy and acts as if they are trustworthy in that way. The behavioral dimension of trust corresponds with certainty in our 3-level basic knowledge structure, which is one of four properties belonging to the class of natural reactions R described in the previous chapter.

Zagzebski's multidimensional conception of self-trust reflects an enriched understanding of rationality that includes affective and behavioral components in addition to the typical cognitive component of belief emphasized in computational theories of rationality and models of trust. Zagzebski [5, p. 12] rightly argues that so long as emotions can be appropriate or inappropriate, and actions can be right or wrong, there is no reason to exclude our emotional dispositions and behavior from the domain of rationality. In the case of the affective dimension of trust, for example, Zagzebski [5, pp. 14–15] argues that one can obsess over the trustworthiness of their own faculties. Even if one believes their own faculties are trustworthy and acts in accordance with this belief, one may still not feel trusting toward them as in the extreme case where one has become too suspicious or skeptical. Similarly, Descartes argues that obsession over the trustworthiness of one's own faculties may lead to irresolution and inaction which is irrational. As Zagzebski [5, p. 16] also rightly argues, full-fledged trust in the reliability of our own faculties may be seen as an intellectual virtue because it puts an end to the process of reflection when decisiveness and action are required. It is rational because excessive reflection is not rational. Rationality is another property of the class of natural reactions R in our 3-level basic knowledge structure discussed in the previous chapter.

5.2 Basic Trust and Full-Fledged Trust

Zagzebski draws a second distinction between basic self-trust and full-fledged self-trust analogous to Sosa's distinction between animal knowledge and reflective knowledge. In the case of basic self-trust, Zagzebski [5, pp. 5–6] argues that we begin in an epistemic state of self-trust that lacks the internalist component of the awareness of risk or vulnerability, epistemically. In the first instance, one is epistemically innocent or naïve. Basic self-trust involves a natural, pre-reflective type of trust in one's own self that is inescapable. The trust of young children, Zagzebski tells us, is an example of this minimal type of trust. The inescapability of basic trust is also a property of the class of natural reactions R in our 3-level basic knowledge structure discussed in the previous chapter.

Zagzebski's conception of basic self-trust is intended as a criticism of those who think of self-trust as the outcome of a sophisticated line of argument or as the product of reflective knowledge. In particular, it is a criticism of Foley's suggestion that basic self-trust is an epistemically inappropriate state to which we retreat when we do not have adequate justification for the reliability of our faculties and opinions (explicit beliefs) taken as a whole. Far from being a fallback position that is indefensible, Zagzebski [5, pp. 8–10] argues that basic self-trust is a necessary condition for knowledge. Moreover, Zagzebski [5, p. 6] argues that basic self-trust is defensible from the vantage point of full-fledged self-trust which involves the awareness of risk or vulnerability and one's acceptance of it. Thus, basic self-trust is not defensible at the level of animal knowledge where it is operative, but at the higher level of reflective knowledge which requires an endorsing perspective on the sources of one's actions and their reliability.

Like Sosa, Zagzebski argues that the reflective person will want to know whether his trust is defensible or justified in accordance with the PC principle stated in the previous chapter. The *prima facie* justification of the reliability of our faculties through basic self-trust can be epistemically enhanced through full-fledged self-trust. Also like Sosa, Zagzebski [5, p. 7] argues that we can think of rationality as doing a better job of what we do naturally, and that we are rational when we do what we do naturally self-reflectively so. For this reason, initial trust in the reliability of our own faculties needs to be defended or better justified. Unlike Sosa, however, this is why Zagzebski [5, p. 19] argues that basic, pre-reflective self-trust is not a virtue, properly speaking. Basic trust gives us access to a sort of knowledge that has minimal justification. To achieve full-fledged trust, this knowledge needs to be enhanced through having an epistemic perspective, whereby we can see that our trust is justified in accordance with the PC principle. Thus, whereas our implicit beliefs or commitments B have *prima facie* justification through the class of natural reactions R, the explicit beliefs or commitments B* formed through reason-based competences such as intellectual virtues enhance this initial justification.

Like basic self-trust, Zagzebski [5, p. 6] argues that basic trust in other people is also inescapable. Thus, basic self-trust and basic trust in other people are both members of the class of natural reactions R in our 3-level basic knowledge structure because they share the properties of inescapability, basicality (being constituted by nature), certainty, and rationality. Like trust in one's own self, trust in other people is the starting point for reflection. If one trusts the generic reliability of one's own faculties, and if one accepts the principle that like cases should be treated alike, Zagzebski [5, p. 16] argues that one is rationally committed to trusting the generic reliability of the faculties of other people. Thus, Zagzebski [5, p. 18] concludes that one ought to hold the same attitude toward the faculties of other people as one holds toward one's own faculties.

Like full-fledged self-trust, which involves the awareness of risk or being vulnerable to falsehoods, Zagzebski [5, p. 6] argues that full-fledged trust in other people involves the awareness of the ways other people can harm us epistemically. Trust in other people is inescapable, irresistible and necessary, moreover, because the epistemic perspective of any single person is limited. Since other people and the community are also bearers of knowledge, Zagzebski [5, p. 23] suggests that the richness and scope of human knowledge transcend their expression in any single epistemic perspective held by an individual. No person can know all there is to know. The awareness of this inherent epistemic limitation elicits the intellectual virtues of humility and open-mindedness. Through the awareness of risk or vulnerability and one's acceptance of it, one acknowledges their own limitations, epistemically, and must trust other people to enlarge and enrich their own epistemic perspective by exercising these intellectual virtues. Thus, enhancing basic trust in other people may be seen as a corrective to excessive basic self-trust.

Zagzebski argues that basic trust in other people depends on basic self-trust. The intellectual virtues of humility and open-mindedness restrain basic self-trust. But these traits of character presuppose basic self-trust in order to restrain it. Neither trait could be a virtue unless we assume it is reasonable to trust our own faculties. Humility derives from one's awareness and acceptance of vulnerability to falsehoods, and one's dependence on other people for attaining truth. As such, humility restrains our level of confidence in the correspondence between our own faculties and the external objects with which they aim to put us in touch. Open-mindedness restrains basic self-trust mostly by enhancing basic trust in other people. This trait requires us to think about an issue from the epistemic perspective or worldview of other people. But, as Zagzebski [5, pp. 20–21] argues, the trait could not be a virtue unless basic trust in other people were *prima facie* justified. Thus, open-mindedness is both an enhancement of basic trust in other people and a restraint on basic trust in one's own self.

Zagzebski's argument that basic trust in other people T* depends on basic self-trust T, however, suggests that T is more epistemically basic than T*. This argument is perhaps motivated by an implicit commitment to internalism, or an attempt to assuage the internalist's criticism of externalist theories of knowledge such as her theory of trust. This implication, whether intentional or not, leads to a contradiction when she also argues that T and T* are both basic, one neither more basic than the other. Thus, T* is not based on T, at least not in the epistemic sense of one type of trust being more basic than the other. This conclusion is more consistent with externalism in its criticism of internalism in assuming the primacy of the self over the community. If both types of trust are equally basic, and they are members of the same class of natural reactions R in our 3-level basic knowledge structure, we need not transfer justification from T (the premise) to T* (the conclusion) as Zagzebski [5, pp. 24–25] explicitly does in the following argument.

P1   Basic self-trust T is reasonable because it is natural and is found upon reflection to be inescapable.

P2   Basic trust in other people T* is also natural and is found upon reflection to be inescapable.

C   Therefore, basic trust in other people T* is reasonable because it is a commitment of consistent self-trust.

Premises P1 and P2, independently held, are consistent with our 3-level basic knowledge structure. Both basic self-trust T and basic trust in other people T* belong to the class of natural reactions R which share the properties of basicality (being constituted by nature), inescapability, certainty, and rationality. All normal adult humans share these two natural reactions. We need not accept premises P1 and P2 jointly held, however, which imply the primacy of the self over the community. If our beliefs, whether implicit or explicit, are rooted in natural reactions or shared agreements in judgment, knowledge is social from the start.

Perhaps a better model for understanding the relation between basic self-trust and basic trust in other people is a mutuality model rather than mutual basing relations. Whereas the second model involves a circle of justification where justification is transferred from premises to conclusions or conversely, a mutuality model shows how one reaction can operate on and moderate the other reaction, either by enhancing it or restraining it through the exercise of intellectual virtues at the level of reflective knowledge. At the level of animal knowledge, basic trust in either of its forms T or T* involves an automatic or spontaneous natural reaction, whereby a ground-level animal competence is exercised. At the level of reflective knowledge, however, these basic trust dispositions or competences can be enhanced or restrained, not in the sense of acquiring or losing an additional measure of justification for them, but in the sense of exercising appropriate intellectual virtues that improve the correspondence between the beliefs formed by them and the world. Thus, consistent with our 3-level basic knowledge structure, Zagzebski [5, p. 24] suggests that intellectual virtues presuppose basic self-trust or basic trust in other people as the case may be, and either enhance or restrain basic trust as the situation demands.

This requirement for knowledge is different from the requirement stated in the PC principle, which implies that basic trust itself needs to be better justified through having an epistemic perspective on one's animal dispositions. Instead, according to the revised version of the principle PC* stated in the previous chapter, the beliefs formed through basic trust in either of its forms need to be better justified through having an epistemic perspective. At the level of animal knowledge, there can be no dialectical relation between these two forms of basic trust. So long as one remains unaware of the risk or vulnerability from either basic trust disposition, epistemically, there can be no relation between them. Basic self-trust and basic trust in other people come into relation only through the development of full-fledged trust at the level of reflective knowledge.

Full-fledged trust involves the awareness of risk or vulnerability and its acceptance in the light of which the intellectual virtues that depend on reason-based competences can improve the correspondence between the beliefs formed through our basic trust dispositions and the world. This claim reflects the fact that one can be either too trusting of one's own self or other people in the absence of the awareness of risk or vulnerability. Or one can be too suspicious or skeptical of one's own self or other people in its presence. To be properly trusting, according to Zagzebski, one needs to exercise intellectual virtues in monitoring, controlling, and moderating these two different natural reactions on a case-by-case basis. This implies that basic self-trust and basic trust in other people are moderating factors in our cognitive structure that need to be balanced at a more reflective level. This balance can be achieved by exercising appropriate intellectual virtues that enhance or restrain our basic trust dispositions to avoid becoming either too suspicious or too trustful. Both extremes are epistemically inappropriate states that impede knowledge. One may either have too much confidence in the reliability of one's own faculties and not enough confidence in those of others. Or one may have too much confidence in the reliability of the faculties of other people and not enough confidence in their own faculties.

Another problem in Zagzebski's theory of trust to which we have already alluded concerns the relation between trust and distrust, or doubt, and whether doubt plays an epistemic role in satisfying the PC principle which states that knowledge is enhanced through justified trust. One would expect that doubt is involved in the awareness of risk or vulnerability and its acceptance as a necessary condition for full-fledged trust. Instead, Zagzebski suggests that trust and doubt are incompatible, though it is unclear from her other writings whether this suggestion is intended. In particular, Zagzebski [5, p. 11] argues that we need to either doubt our beliefs and lose trust in the faculties that produced them, or we need to trust in a fully reflective way. The first part of the disjunction implies, at least on the surface, that trust and doubt are mutually exclusive. To doubt our beliefs means that we must lose trust in the faculties that produced them. The second part of the disjunction, however, implies that in order to trust in a fully reflective way one requires an epistemic perspective, whereby one can see that his trust in the reliability of his animal faculties is justified. But if trust and distrust cannot co-exist, and if distrust or doubt depends on reflective knowledge, then trust and reflective knowledge cannot co-exist and we can never satisfy the PC principle which is precisely what the skeptic claims. Thus, in order to refute the skeptic, we need to show that trust and distrust can co-exist.

Sosa's distinction between animal knowledge and reflective knowledge might help to resolve this debacle. Whereas trust and distrust or doubt are mutually exclusive at the level of animal knowledge, they can co-exist and indeed are required to do so at the level of reflective knowledge. The first claim is consistent with our 3-level basic knowledge structure, according to which basic trust is a member of the class of natural reactions R whose properties include certainty, the opposite of uncertainty or doubt. The second claim is also consistent with our 3-level basic knowledge structure since it helps to explain the qualitative difference, epistemically, between animal knowledge and reflective knowledge. At the level of reflective knowledge, doubt is not only compatible with trust, it is necessary to help us trust in a fully reflective way. Doubt is an integral component of the awareness of risk or vulnerability and its acceptance in full-fledged trust. Far from being mutually exclusive, trust and doubt are complementary cognitive processes or competences necessary for knowledge at a more reflective level. Trusting the reliability of our animal faculties, and then doubting them as the evidence demands in the light of a belief that has not been well formed, are what make reflective knowledge and justified trust a higher and more desirable state of knowledge in contrast to basic, pre-reflective trust at the level of animal knowledge.

Despite these minor problems, Zagzebski helps us see that rationality depends on two different but complementary basic trust dispositions or natural reactions. Rationality, moreover, involves acts of intellectual virtue in monitoring, controlling, and moderating the epistemic influences and interaction of these basic trust dispositions. This interactive view of rationality is more commensurate with a virtue approach to knowledge, which sees knowledge as something we actively acquire and manifest in a performance that can be measured rather than as something we passively receive by brute luck. Not only is rationality rooted in basic trust dispositions or natural reactions that are connected to our instinctual rational conduct at the animal level. Acts of epistemic virtue that are connected to our rational conduct at a more reflective level play a central role in guiding these foundational and ground-level dispositions or competences in achieving socially valuable ends such as cooperation and collaboration based on epistemically appropriate states, while avoiding socially undesirable ends such as selfishness and conflict that result from epistemically inappropriate states.

5.3 Epistemically Rational Belief and Responsible Belief

If our 3-level basic knowledge structure is to be socially relevant, we need to understand how it can be generalized from individual cognitive structures to groups or systems. Sosa [3, p. 52] suggests how we might proceed in arguing that the conception of an individual competence can be generalized to a social competence exercised by a group collectively. In previous chapters, we have seen how implicit beliefs or commitments are acquired through animal dispositions or competences such as perception, which depend in turn on a distinctive type of animal competence. Whereas Sosa calls this competence apt belief, Zagzebski calls it basic trust. Like Zagzebski, Sosa argues that the mechanisms of silent enculturation at the level of animal knowledge exert their influence non-rationally rather than through the operation of reasons. Non-rational does not mean that these mechanisms are irrational, only that they are not based on reasons that we are aware of as motivating factors in the formation of our implicit beliefs or commitments as required by internalism.

Sosa [3, p. 136] argues that one mechanism of silent enculturation is our natural disposition to trust or receive testimony from other people, which is one of the most common sources or bases of our knowledge. Most of what we know about the world beyond the scope of personal experience derives from the oral or written say-so of other people. Sosa [3, p. 93] argues that testimony is a disposition or competence built into our first nature, whereby we readily trust or receive the say-so of other people when we hear it or read it. Since testimony reliably gives access to animal knowledge, it is an instance of apt belief. According to Sosa [3, pp. 94–97], we can think of this competence as seated not only in the cognitive structure of an individual, but in a group whose exercise of it leads through testimonial links to the correctness of one's present belief. Thus, testimonial knowledge can take the form of a belief whose correctness is attributable to a complex social competence that is seated only partially in the individual. The individual shares credit for the correctness of the belief with the group. Zagzebski extends Sosa's suggestion in arguing that our innate ability to receive testimony can be directed toward inanimate objects as well as other people. If this is true, then our basic trust in other people manifested by the competence of testimony can be directed toward systems.

To justify our initial trust in the reliability of testimony in accordance with the PC principle, whether the testimony is received from other people or from systems, we need full-fledged trust which requires the awareness of risk or vulnerability and its acceptance. But, as Foley [6] argues, we also need responsible belief which requires that we base our decisions on beliefs that are good enough given a context, since our time and intellectual or computational resources are limited by practical constraints. According to Foley [6, p. 229], our everyday evaluations of beliefs and actions are saturated with reasons, and presuppose notions of rationality, reasonableness, or one of their cognates. These notions, however, leave us within a circle of terms for which we need a satisfactory philosophical account. Like Sosa and Zagzebski, Foley argues that we acquire most of our beliefs with no or little reflection. One is naturally inclined or disposed to believe, for example, that there is a red wall because one sees it. There is no deliberation about whether to trust the reliability of our animal faculty of perception. Such belief-forming and belief-sustaining dispositions are part of our instinctual rational conduct, and thus are reliable and trustworthy ways to acquire beliefs absent some indication to the contrary. Foley [6, p. 224] further argues that our rational conduct operates on automatic pilot until we have reason to suspect that a belief was not well formed, and that we need to make adjustments in our cognitive doings at a more reflective level. Since most cognitive processing is conducted in a largely automatic fashion, Foley agrees with Sosa and Zagzebski that it is important to cultivate intellectual virtues that help us do a better job of what we do naturally.

Also like Sosa and Zagzebski, Foley defines an intellectual virtue as a trait, skill, habit, or reason-based competence. Unlike Sosa, however, whose interest in the intellectual virtues is admittedly instrumental in helping us understand epistemic normativity, Foley and Zagzebski are interested in applying the conception of an intellectual virtue in a pragmatic and social context where the emphasis is on practical rather than theoretical rationality. According to both philosophers, intellectual virtues can be seen not only as reason-based competences that are conducive to the intellectual goal of now having accurate and comprehensively coherent beliefs. They can also be seen as reason-based competences that are conducive to the pragmatic goal of achieving epistemically appropriate states that facilitate socially valuable ends such as cooperation and collaboration. Thus, whereas intellectual virtues such as judiciousness, conscientiousness, and rigor are traits that promote one's having accurate and comprehensively coherent beliefs, intellectual virtues such as humility, open-mindedness, and fairness are traits that promote epistemically appropriate states that facilitate cooperation and collaboration.

What is needed, Foley argues, is a general theory of rationality that includes both types of goals or ends, pragmatic as well as intellectual. In response, Foley proposes the following general schema which provides a sufficient condition for rationality, and defines a rational criterion for choosing between plans that aim to bring about one or other of these goals or ends.

S   A plan (decision, action, strategy, belief, etc.) is rational in sense X for an
     individual if it is epistemically rational for the individual to believe that it
     would acceptably contribute to his or her goals of type X.

Observe that the general schema includes actions as well as beliefs which is consistent with a more substantive interpretation of the analogy between virtue epistemology and virtue ethics with their respective emphases on beliefs and actions.
The notion of epistemic rationality or normativity serves as a theoretical anchor for
other notions of rationality. According to Foley [6, p. 217], the notion of epistemic
rationality escapes circularity because no matter what account of rational belief is
offered, whether it be foundationalist, coherentist or reliabilist, it need make no
reference to any other notion of rationality than epistemic rationality.

   In principle, the goals of a plan can include whatever has value for an individual
or group. A plan acceptably contributes to a goal if the estimated desirability of the
plan is sufficiently high, where estimated desirability is a function both of what it is
rational to believe about the probable effectiveness of the plan in promoting one's
goal and the relative value of other goals that could have been selected. The
estimated desirability of the plan, moreover, must be sufficiently high given a
context, where the context is determined by the relative desirability of other
available options and their relative accessibility. Thus, the fewer alternative options
there are with greater estimated desirability, the more likely it is that the plan is
rational. If there are alternative options with greater estimated desirability, but they
are only marginally superior or cannot be readily implemented, the plan is all the
more likely to be rational. Foley [6, p. 216] argues that it is rational because it is
good enough given a context. This qualification allows us to adjust how much time
and effort it is reasonable to spend on pursuing a goal given a context.

   Although the general schema permits any number of definitions of rationality, Foley [6, p. 223] suggests that we can define two varieties of rationality as
instances of S, one based on the intellectual goal of now having accurate and
comprehensively coherent beliefs, and the other based on the pragmatic goal of
achieving socially valuable ends such as cooperation and collaboration by promoting
epistemically appropriate states.

D1   Believing p is epistemically rational if it is epistemically rational for one to
     believe that believing p would acceptably contribute to the intellectual goal of
     one's now having accurate and comprehensively coherent beliefs.
D2   One responsibly believes p if one believes p and one also has an epistemically
     rational belief that the processes by which one has acquired and sustained the
     belief p have been acceptable given the limitations on one's time and
     capacities and given all of one's goals, pragmatic as well as intellectual and
     long-term as well as short-term.

Definition D2 entails definition D1 which serves to theoretically anchor the notion of
responsible belief in the more restrictive and rigorous notion of epistemic rationality
or normativity. Thus, D2 provides a more precise definition of sufficient or practical
rationality, in contrast to the way the term is used rather loosely in social science and

computer science to refer to any number of different notions of rationality. At the same time, definition D2 extends the scope of definition D1 in allowing us to apply the more rigorous notion of epistemic rationality or normativity in a social context.

These definitions, moreover, are complementary. Whereas definition D1 states that rational decision-making requires accurate and comprehensively coherent beliefs at the time the decision is made, definition D2 places practical limitations on the attainment of this intellectual goal in the light of one's available time and intellectual or computational resources in achieving some pragmatic goal. In the case of autonomous rational agents in computer networks, D2 places practical limitations on D1 in the light of acceptable time and space complexity constraints.

The definition of responsible belief in terms of the notion of epistemic rationality or normativity (epistemically rational belief) applies only to the evaluation of implicit and explicit beliefs at the level of reflective knowledge. To make this clear, Foley draws a distinction between responsible belief and non-negligent or innocent belief which illuminates Sosa's distinction between animal knowledge and reflective knowledge. With respect to a belief that has been acquired automatically and non-voluntarily without deliberation, Foley argues that the belief is responsible if one has an epistemically rational belief that the faculties and skills that formed the belief are acceptable in the light of all of one's goals. This corresponds to the requirement of an endorsing perspective at the level of reflective knowledge in Sosa's virtue perspectivism, whereby one can see the reliability or acceptability of one's animal competences such as perception, albeit absent the assumption of unlimited time and intellectual or computational resources.

Like Sosa's criticism of internalism, however, Foley [6, p. 223] argues that we often do not have a good idea of how we came to believe what we do. We may not remember or never know the justificatory conditions that were in play at the moment of acceptance. Thus, with respect to most of our beliefs, we may not believe that the processes that led to them were acceptable or reliable. But we also need not believe or have reasons for believing that these processes were unacceptable or unreliable. Thus, Foley denotes the class of implicit beliefs or commitments formed by animal competences as non-negligent because the definition of responsible belief does not apply at the level of animal knowledge. Responsible belief requires awareness and conscious control of our animal beliefs as in internalism. Non-negligent belief corresponds to Sosa's definition of animal knowledge as apt belief. Our animal competences such as perception are apt because they reliably form beliefs that are preponderantly true, even if occasionally false. In the next chapter, we will examine how Foley's definition of responsible belief can help us integrate Sosa's account of epistemic rationality or normativity with Zagzebski's conception of full-fledged trust in a unified theory of trust that can be applied in a social context.

References

1. Zagzebski L (2001) Must knowers be agents? In: Fairweather A, Zagzebski L (eds) Virtue epistemology: essays on epistemic virtue and responsibility. Oxford University Press, Oxford, pp 142–157
2. Korsgaard CM (1996) Sources of normativity. Cambridge University Press, Cambridge
3. Sosa E (2007) Apt belief and reflective knowledge. Vol. 1: A virtue epistemology. Oxford University Press, New York
4. Sosa E (2009) Apt belief and reflective knowledge. Vol. 2: Reflective knowledge. Oxford University Press, New York
5. Zagzebski L (2014) Trust. In: Tempte K, Boyd C (eds) Vices and their virtues. Oxford University Press, New York (in press), pp. 269–284. Much of Sects. 2–3 in the draft copy are based on Chap. 2 of the following book by the same author (2012) Epistemic authority: a theory of trust, authority, and autonomy in belief. Oxford University Press, New York (in press). All references are to the draft copy: http://www.ou.edu/ouphil/faculty/zagzebski/Trust.pdf. Accessed 15 Jul 2014
6. Foley R (2012) The foundational role of epistemology in a general theory of rationality. In: Fairweather A, Zagzebski L (eds) Epistemic authority: essays on epistemic virtue and responsibility. Oxford University Press, New York, pp 214–230

Chapter 6
A Normative Virtue-Based Trust Model

We have discussed how our 3-level basic knowledge structure can be integrated with other theories of trust to yield a more enlightened perspective on the relation between trust and rationality. We began with Zagzebski's theory of trust which draws a distinction between self-trust and trust in other people and another distinction between basic trust and full-fledged trust. We saw how Zagzebski employs these distinctions in developing the notion of being properly trusting. We proceeded to discuss Foley's definition of responsible belief which allows us to develop a non-circular theory of rationality by anchoring the definition in the more restrictive and rigorous notion of epistemic rationality or normativity. Foley's definition of responsible belief also allows us to place practical limitations on the time and intellectual or computational resources required to pursue the intellectual goals of now having accurate and comprehensively coherent beliefs and full-fledged trust given a context.

In this chapter, we discuss in more detail how Foley's definition of responsible belief can be used to integrate Sosa's account of epistemic normativity, or epistemically rational belief, with Zagzebski's conception of full-fledged trust in a unified theory of trust. Although Foley helps us see how a non-circular theory of rationality can be derived from a virtue approach to knowledge such as Sosa's virtue perspectivism, we need to understand how intellectual virtues are related to this general theory of rationality. In particular, we need to understand how intellectual virtues are related to the basic trust dispositions in Zagzebski's theory of trust, which we want to monitor, control, and moderate through responsible belief. This unified theory of trust, with appropriate modifications from our 3-level basic knowledge structure, can be translated into a practical, virtue-based trust model that can be applied in a social context. The trust model helps us understand how we can achieve socially valuable ends such as cooperation and collaboration, while avoiding socially undesirable ends such as selfishness and conflict, through epistemically appropriate states and the exercise of intellectual virtues associated with them.

Finally, we need to understand how a virtue-based trust model can be extended to an intercultural context. Since most societies place a high value on virtue and character as traits of trustworthiness, a virtue-based approach to trust seems well

© The Author(s) 2014
M.G. Harvey, *Wireless Next Generation Networks*,
SpringerBriefs in Electrical and Computer Engineering,
DOI 10.1007/978-3-319-11903-8_6

suited to the task of defining an intercultural theory of trust. The concept of a social exchange trust policy is introduced, whereby the awareness of risk or vulnerability and its acceptance is distributed more or less equally between both entities in the trust relation based on the moderating notion of balance, symmetry, or harmony. By integrating the Western conception of balance or symmetry with the Eastern conception of harmony, we can develop a cross-cultural approach to computer-mediated interactions based on the idea of achieving symmetry or harmony between both sides in a trust relation through mutual adjustment, in contrast to one side dominating the other side through argument and persuasion.

6.1 Trust Relations, Epistemic States, and Social Ends

We begin with a sketch of a virtue-based trust model that shows how the general theory of rationality we have been proposing might be applied in a social context. This trust model helps us understand how to achieve the intellectual and pragmatic goals of promoting epistemically appropriate states that facilitate socially valuable ends such as cooperation and collaboration. At the same time, the trust model can help us understand how to avoid epistemically inappropriate states that lead to socially undesirable ends such as selfishness and conflict. Table 6.1 lists four fundamental epistemic states and their corresponding social ends, if any, and shows how each state can be expressed as a function of two basic trust dispositions, self-trust T and trust in other people T*. Each epistemic state is associated with either a null end, a politico-economic end, or a socio-cultural end. Politico-economic ends include selfishness, a socially undesirable end, and cooperation, a socially valuable end. Similarly, socio-cultural ends include conflict, a more severe socially undesirable end, and collaboration, a more ideal socially valuable end.

The trust model assumes that intellectual virtues are required at the level of reflective knowledge to monitor, control, and moderate the relative influences of basic self-trust and basic trust in other people operative at the level of animal knowledge in the achievement or avoidance of one or more of these epistemic states

Table 6.1 Dependence of epistemic states on T and T*

Self-trust T	Trust in others T*	Epistemic level	Epistemic state	Politico-economic ends	Socio-cultural ends
0	0	Pre-reflective	Basic trust	Ø	Ø
0	1	Unreflective	Too trusting	Ø	Ø
1	0	Excessively reflective	Too suspicious	Selfishness	Conflict
1	1	Reflective	Full-fledged trust	Cooperation	Collaboration

and their corresponding social ends. Although the trust model is flexible in that other social ends can be substituted for the ones listed in Table 6.1, the social ends chosen seem well suited for a practical trust model. The epistemic states range from basic trust to full-fledged trust in Zagzebski's [1] theory of trust. Whereas two epistemic states (00/01) are pre-reflective and unreflective in that they involve no or little awareness of risk or vulnerability and its acceptance respectively, the remaining two epistemic states (10/11) are reflective in that they meet this internalist requirement.

These four fundamental epistemic states can be further distinguished by which form of basic trust, T or T*, dominates the trust relation. Typically, one form of basic trust is more operative and influential than the other at the animal level in a given situation. If T and T* exert the same influence on the formation of our beliefs, then an increase or enhancement in the level of one type of basic trust should correspond to a decrease or restraint in the level of the other type of basic trust. Thus, we can initially think of the relation between basic self-trust and basic trust in other people as one of inverse proportionality where $T = 1/T^*$. But if one basic trust disposition exerts more influence on the formation of our beliefs than the other so that one of them is more dominant in the trust relation, we need a constant of proportionality C that depends on the awareness of risk or vulnerability V in a given epistemic situation to maintain their balance.

TR   For a given epistemic situation where one is aware of risk or vulnerability V and accepts it, and either T or T* is the dominant factor in the trust relation such that $T \neq 1/T^*$, there exists some constant of proportionality C that depends on V so that $TC = 1/T^*$.

This trust relation states that the relation between basic self-trust and basic trust in other people should ideally be symmetric. If it is not symmetric, then the trust relation needs to be adjusted or moderated according to a scalar value that depends on the awareness of risk or vulnerability and its acceptance to maintain balance between the two basic trust dispositions. Thus, to be properly trusting in a given epistemic situation means that an individual cognitive structure or system is in a state of equilibrium. The trust relation is based on Zagzebski's [1] conception of being properly trusting and Foley's [2] definition of responsible belief. The latter places practical limitations on the time and intellectual or computational resources required to pursue Zagzebski's intellectual goal of full-fledged trust and Sosa's [3, 4] intellectual goal of epistemically rational belief or now having accurate and comprehensively coherent beliefs.

Table 6.1 shows two extremes. At one extreme, we have state 00. If neither T nor T* is enhanced, one is in a start state of basic, pre-reflective trust that lacks the awareness of risk or vulnerability and its acceptance. In this pre-reflective or animal state, there are no socially valuable ends that can be achieved or socially undesirable ends that can be avoided, since there can be no intellectual virtues at a pre-reflective level to monitor, control, and moderate the relative influences of T and T* in the achievement or avoidance of social ends. Recall the argument made in the previous chapter that in exercising an intellectual virtue one is exercising a power that requires agency, and there can be no conscious agency at a pre-reflective level without introducing circularity in our theory of rationality. At the other extreme, we have state 11. If both T and T* are enhanced in direct proportion to each other according to the trust relation TR, one is in a goal state of full-fledged trust which involves the awareness of risk or vulnerability and its acceptance. This state facilitates the socially valuable ends of cooperation and collaboration as basic trust in other people T* is enhanced, while basic self-trust T is restrained relative to some constant of proportionality that depends on one's awareness of risk or vulnerability and its acceptance in the trust relation. Intellectual virtues that tend to bring about cooperation and collaboration include humility, open-mindedness, and fairness.

There are two cases in between these extremes which are more typical of our actual epistemic situation. In the case of state 01, if T* is maximally enhanced while T is maximally restrained in violation of the trust relation TR, one tends toward the epistemically inappropriate state of being too trusting. Since intellectual virtues are downplayed or suppressed, this state degenerates into cognitive quietism which is effectively pre-reflective, marked by no or little awareness of risk or vulnerability and its acceptance in the trust relation. Thus, as in the case of basic trust at the level of animal knowledge, there are no socio-cultural ends or politico-economic ends that can be achieved or avoided, since there can be no intellectual virtues that enhance or restrain basic trust in either of its forms to bring about these social ends. Conversely, in the case of state 10, if T is maximally enhanced while T* is maximally restrained, again in violation of the trust relation TR, one tends toward the epistemically inappropriate state of being too suspicious. Although reflective, this state degenerates into absolute autonomy or total self-reliance which precludes the awareness of risk or vulnerability and its acceptance. In this state, one is convinced that one cannot be wrong or make mistakes, epistemically. In contrast to full-fledged trust, this state leads to the socially undesirable ends of selfishness and conflict, the contraries of cooperation and collaboration respectively.

From Table 6.1 we can derive an epistemic state diagram and a graph of relations between the social ends associated with these epistemic states. Observe that the social ends shown in Fig. 6.1 appear only along the reflective and excessively reflective dimensions of the epistemic state diagram in Fig. 6.2, since the achievement or avoidance of social ends depends on the exercise of appropriate intellectual virtues that depend in turn on reason-based dispositions or competences. Thus, only null ends appear along the pre-reflective and unreflective dimensions of the epistemic state diagram. Although these epistemic states might still have social ends that are brought about by brute luck, they are not brought about reflectively so through responsible belief and the exercise of reason-based competences such as intellectual virtues. The social ends that appear along the reflective and excessively reflective dimensions of the epistemic state diagram are either social ends we want

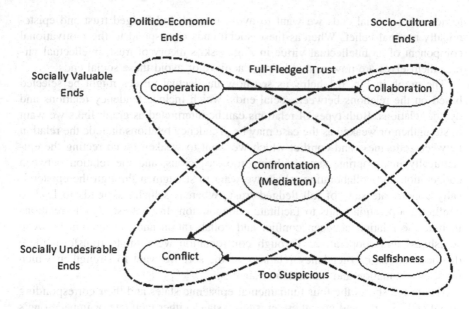

Fig. 6.1 Graph of relations between social ends

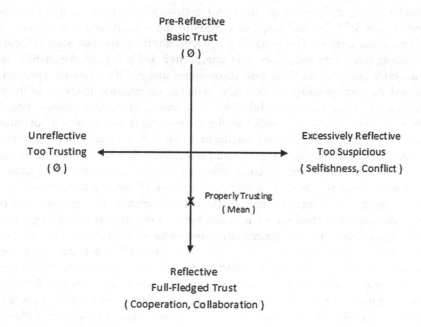

Fig. 6.2 Epistemic state diagram

to achieve or social ends we want to avoid through full-fledged trust and episte-mically rational belief. Whereas these social ends correspond to the motivational component of an intellectual virtue in Zagzebski's theory of trust, intellectual vir-tues themselves are powers that help us achieve or avoid these social ends.

The graph in Fig. 6.1 helps us see how intellectual virtues might be selected based on the relations between social ends, which include tendency relations and dyadic relations. Both types of relations can be interpreted as graph links we want to strengthen or weaken as the case may be. Tendency relations include the relation between selfishness and conflict which we want to weaken by correcting the epi-stemically inappropriate state of being too suspicious, and the relation between cooperation and collaboration which we want to strengthen through the epistemi-cally appropriate state of full-fledged trust. Whereas selfishness tends to lead to conflict, cooperation tends to facilitate collaboration. In contrast, dyadic relations include the relation between conflict and collaboration and the relation between selfishness and cooperation. Through confrontation we want to transform these dyadic relations into tendency relations of the sort we want to strengthen which requires negotiation and mediation strategies.

Figure 6.2 shows the four fundamental epistemic states and their corresponding social ends, if any, and how these epistemic states either facilitate or impede one's progress toward the goal state of full-fledged trust and epistemically rational belief. Whereas the epistemically appropriate state of full-fledged trust facilitates the socially valuable ends of cooperation and collaboration, the epistemically inap-propriate state of being too suspicious leads to the socially undesirable ends of selfishness and conflict. The remaining epistemically inappropriate state of being too trusting has a null end, since it is unreflective and precludes the intellectual virtues from operating on basic trust dispositions that could change the epistemic state and its corresponding social ends. Whereas cooperation tends to facilitate collaboration in the case of full-fledged trust and epistemically rational belief, selfishness tends to lead to conflict in the case of being too suspicious of other people. In the latter case, the very intellectual virtues that enhance basic self-trust such as intellectual courage, firmness, and perseverance degenerate into intellectual vices when unaccompanied by complementary intellectual virtues that enhance basic trust in other people such as humility, open-mindedness, and fairness.

By analogy with nature, we may think of the interaction of these two sets of complementary intellectual virtues as a combination of forces and restoring forces that act against each other to dynamically maintain an individual cognitive structure or system in equilibrium. Thus, in contrast to being either too trusting (over-confident) or too suspicious (over-diffident) of other people or entities, full-fledged trust is being properly trusting because it reliably guides an epistemically well-functioning human being in achieving a socially valuable end or avoiding a socially undesirable end. This argument is nothing more than an epistemic version of Plato's [5] moral advice about how one should conduct a virtuous life that will lead to the goal of happiness: Let one know how to choose the mean and avoid the extremes on either side as far as possible.

In general, we want to select an appropriate mix of intellectual virtues that facilitate the socially valuable ends of cooperation and collaboration, while avoiding the socially undesirable ends of selfishness and conflict. To achieve cooperation and ultimately collaboration, we need to avoid selfishness and conflict by selecting intellectual virtues that enhance basic trust in other people while restraining excessive basic self-trust. At the same time, in order to avoid being too trusting of other people, we need to select intellectual virtues that enhance basic self-trust while restraining excessive basic trust in other people. To be properly trusting, we need to progress toward the goal state of full-fledged trust and epistemically rational belief. To do so requires that we strike a balance between these two sets of intellectual virtues so that they complement each other rather than cancel each other. Table 6.2 lists two sets of complementary intellectual virtues. Obviously the list is not exhaustive, but it is sufficient for illustration. Suppose that one exercises humility to enhance basic trust in the judgments of other people. One also needs to exercise intellectual courage to avoid being too trusting of those judgments. Conversely, suppose that one exercises intellectual firmness to enhance basic trust in one's own judgments. One also needs to exercise open-mindedness to avoid being too suspicious of the judgments of other people which could show where one is mistaken.

This virtue-based trust model can be adapted across cultures in that intellectual virtues can be selected in accordance with their relative importance in a given society in the light of its functional and social needs. However, the plan or strategy (in Foley's sense of the term) of selecting pairs of complementary intellectual virtues that help balance the epistemic influences of the basic trust dispositions remains the same. In this way we can define a practical trust model, at least in rough outline, whose goal is to maximize cooperation and collaboration and minimize selfishness and conflict through mutual adjustment, moderation, and self-control. This pragmatic goal can be achieved by moderately enhancing basic trust in other people while restraining excessive basic self-trust. At the same time, entities need to be protected from accepting too much risk or vulnerability in the trust relation by moderately enhancing basic self-trust while restraining excessive basic trust in other people.

Table 6.2 Two sets of complementary intellectual virtues

Intellectual virtues conducive to cooperation and collaboration	Intellectual virtues conducive to being properly trusting
Humility	Intellectual courage
Open-mindedness	Firmness
Fairness	Perseverance

6.2 A Unified Theory of Trust

Now we need to show how the contributions of Sosa, Zagzebski, and Foley can be
integrated in a unified theory of trust that can be translated into a practical trust model
such as the one sketched above. These contributions allow us to refine our 3-level
basic knowledge structure, as well as to show how this individual cognitive structure
can be generalized to a social context. Figure 6.3 shows the basic trust feedback loop
for an individual epistemology such as our 3-level basic knowledge structure. The
process diagram shows how dispositions or competences at the level of animal
knowledge K such as perception depend on a distinctive type of animal competence
called basic trust which operates at a pre-reflective or non-rational epistemic level.
Recall from a previous chapter that a clear distinction between pre-reflective epistemic states (reactions) and unreflective epistemic states (implicit beliefs) is required
to avoid vicious circularity in our general theory of rationality. This pre-reflective,
animal level corresponds to the bottom layer in our 3-level basic knowledge

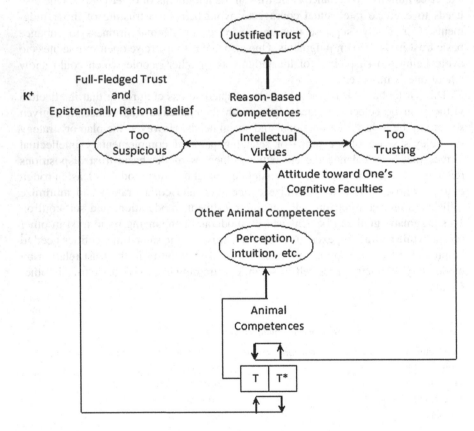

K   Basic Trust

Fig. 6.3 Basic trust feedback loop for a basic knowledge structure

structure, comprised of the class of natural reactions R with its properties of basicality (being constituted by nature), inescapability, certainty, and rationality. This class includes two basic trust dispositions, self-trust T and trust in other people T*.

The process diagram also shows how intellectual virtues, which depend on reason-based dispositions or competences such as deduction and abduction, are necessary to meet the requirement of an endorsing perspective for reflective knowledge K^+ in Sosa's virtue perspectivism and Zagzebski's theory of trust. Whereas deduction serves the analytic function of reason in transferring justification from premises to conclusions through mutual basing relations, abduction serves the synthetic function of reason in combining the cross-level coherence of these belief relations in a more comprehensively coherent epistemic perspective. Through having an endorsing perspective, we are able to satisfy the PC principle which states that knowledge is enhanced through justified trust in the reliability of our animal dispositions or competences. This goal state corresponds to Zagzebski's conception of full-fledged trust and to Foley's conception of epistemically rational belief which is satisfied by Sosa's account of epistemic rationality or normativity. According to the revised version of the principle PC* discussed in a previous chapter, however, it is not the trust disposition itself that needs to be justified but rather the beliefs formed by this disposition, namely, the implicit beliefs or commitments that comprise most of our cognitive structure.

At the higher and more desirable level of reflective knowledge, we are able to discern whether we have been too trusting of the reliability of our own faculties in the formation of our beliefs or too suspicious of their reliability. Through the exercise of relevant intellectual virtues, we are able to maintain a balance between these epistemically inappropriate states. This ideal epistemic state, which involves choosing the mean and avoiding the extremes on either side as far as possible, corresponds to Zagzebski's conception of being properly trusting which facilitates the goal state of full-fledged trust and epistemically rational belief. In particular, for a given epistemic situation, we can enhance or restrain the basic trust dispositions that motivate and influence our rational conduct. Although our rational conduct is instinctual at the level of animal knowledge, we are able to monitor, control, and moderate it at the level of reflective knowledge by taking appropriate corrective actions with respect to the basic trust dispositions that motivate and influence the implicit beliefs or commitments formed by them. The iterative nature of the basic trust feedback loop enables intellectual virtues to operate on the basic trust dispositions with the goal of adjusting or moderating their relative influences on the formation of our beliefs in a given epistemic situation.

In the case of being too trusting of the reliability of our own faculties, we need to select intellectual virtues that enhance basic trust in the reliability of the faculties of other people T* while restraining excessive basic trust in the reliability of our own faculties T. This mutual adjustment and moderation of the basic trust dispositions protects us from having too much confidence in our own cognitive faculties, and requires the awareness of risk or vulnerability and its acceptance. Conversely, in the case of being too suspicious of the reliability of our own faculties, we need to select intellectual virtues that enhance basic trust in the reliability of our own faculties T while restraining excessive basic trust in the reliability of the faculties of other people T*. This mutual adjustment and moderation of the basic trust dispositions protects us from having too little confidence in our own cognitive faculties and from being overly skeptical about their correct operation, and requires autonomy and agency.

We can generalize the process diagram for individual cognitive structures to groups and systems. Figure 6.4 shows the basic trust feedback loop for a social epistemology in contrast to an individual epistemology. Since the emphasis is now on our natural reactions or attitudes toward other people or systems rather than toward our own cognitive faculties, the mutual adjustment and moderation of the basic trust dispositions discussed above is inverted in the process diagram. In the case of being too trusting of other people or systems, we need to select intellectual virtues that enhance basic trust in our own rational abilities T while restraining excessive basic trust in other people or systems T*. This mutual adjustment and moderation of the basic trust dispositions protects us from having too much confidence in the testimony or say-so of other people or systems, and thereby protects entities from the risk or vulnerability associated with false data, beliefs, or information. In wired computer networks, for example, these intellectual virtues are embodied and encoded in centralized security policies, mechanisms, and controls that aim to provide this protection. Conversely, in the case of being too suspicious of other people or systems, we need to select intellectual virtues that enhance basic

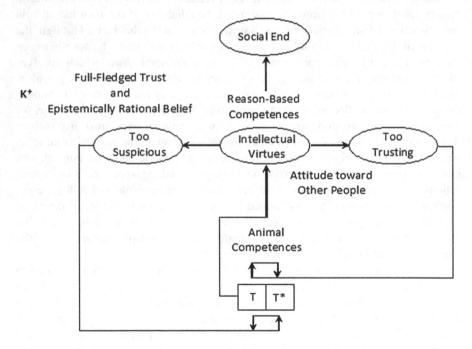

K   Basic Trust

Fig. 6.4 Basic trust feedback loop for a social epistemology

trust in other people or systems T* while restraining excessive basic trust in our own rational abilities T. This mutual adjustment and moderation of the basic trust dispositions protects us from being too skeptical toward other people or systems, thereby facilitating the socially valuable ends of cooperation and collaboration while avoiding the socially undesirable ends of selfishness and conflict.

The intellectual goal of now having accurate and comprehensively coherent beliefs in Sosa's virtue perspectivism, as well as the intellectual goal of full-fledged trust in Zagzebski's theory of trust, assume unlimited time and intellectual or computational resources on the part of the subject or agent in pursuing these intellectual goals. This assumption poses a challenge for practical rationality in a social context, where both time and intellectual or computational resources are often severely limited. This means that the basic trust feedback loop should be able to monitor, control, and moderate the basic trust dispositions subject to real-time constraints. It also means that limitations need to be placed on the time and space complexity requirements of algorithms that must operate on these basic trust dispositions. Thus, we need a definition of sufficient or practical rationality that places appropriate limitations on these intellectual goals to expedite the tasks of gathering and evaluating information to make apt security decisions.

Figure 6.5 shows how Foley's definition of responsible belief can provide a definition of sufficient rationality that places practical limitations on the intellectual goals of now having accurate and comprehensively coherent beliefs and full-fledged trust. Observe that the function diverges to ∞ as both the level of awareness of risk or vulnerability and the level of reflection in the trust relation are increased. Whereas the vertical dimension of the awareness of risk or vulnerability is controlled by Zagzebski's intellectual goal of full-fledged trust, the horizontal dimension of reflection is controlled by Sosa's intellectual goal of epistemically rational belief or now having accurate and comprehensively coherent beliefs. Thus, whereas the vertical axis ranges from basic trust to full-fledged trust, the horizontal axis ranges from animal knowledge to reflective knowledge. The horizontal dimension of reflection also ranges between two epistemically inappropriate states–either being too trusting of other entities (not reflective enough) or being too suspicious of other entities (excessively reflective).

The asymptotic nature of the function reflects the claim made both by Sosa and Zagzebski that excessive concern and reflection over the reliability of one's sources of knowledge are not rational. Thus, we need to draw a boundary at some point x beyond which it is not rational to be concerned over the reliability of one's sources of knowledge. This somewhat arbitrary boundary is indicated by the dashed vertical line or asymptote. This line corresponds to Zagzebski's conception of being properly trusting which is itself a theoretical ideal meant to be approximated by our practical rational conduct at a more reflective level. The point x* marks a region just short of this line by some arbitrarily small value ε that defines a notion of sufficient rationality.

We are now in a position to state how the contributions of Sosa, Zagzebski, and Foley can be integrated in a unified theory of trust that can be translated into a practical trust model.

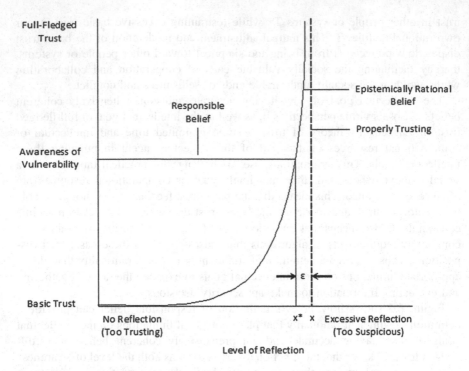

Fig. 6.5 Responsible belief as a functional constraint on intellectual goals

UT   Responsible belief is epistemically rational belief and full-fledged trust that is good enough given a context, where the context determines what is epistemically rational and acceptable to believe for achieving some socially valuable end or avoiding some socially undesirable end.

Responsible belief defines a degree or quality of reflection x* just short of the theoretical ideals of epistemically rational belief and full-fledged trust that is sufficiently rational or good enough given a context, where the context places practical limitations on both intellectual goals in achieving socially valuable ends such as cooperation and collaboration and avoiding socially undesirable ends such as selfishness and conflict. Thus, the value x* defines how much time and intellectual or computational resources it is reasonable to allocate to pursuing the intellectual goals of now having accurate and comprehensively coherent beliefs and full-fledged trust in achieving or avoiding social ends.

6.3 Toward an Intercultural Theory of Trust

We have seen how our 3-level basic knowledge structure can be extended to a social context, where it can be translated into a practical, virtue-based trust model. The question now is whether this trust model can be further extended to an

intercultural context. Since cultures are presently interconnected in a global econ-
omy through computer networks, the question effectively becomes whether the trust
model can successfully function in an open, dynamic environment such as the
mobile Internet. In contrast to the one-way nature of computer-mediated interactions in early computer networks, computer-mediated interactions now involve the
mutual exchange of personal knowledge such as digital credentials and access
control policies. Trust decisions are based on digital credentials and the identity of
their issuer. As wireless next generation networks have stretched the traditional
network boundary, current approaches to trust have come to rely on computational
models. Such trust models, however, are often infeasible or difficult to implement in
the highly contextualized and nuanced environment of the mobile Internet. Mobile
devices cannot handle the large computation and communication overhead due to
their limited resources.

To address the need for a more efficient trust model, Ryutov [6, pp. 8–9] argues
that next generation approaches to trust will have to model the social exchange
between people in an open and dynamic environment. To address the need for a
more flexible trust model, Ryutov argues that a framework based on trust metrics
and cost and utility parameters will have to be developed that can generate ad hoc
trust policies for given social interactions. This argument motivates the following
definition of a social exchange trust policy.

TP   A social exchange trust policy aims to achieve a balance between opposites
      through a mutual adjustment between oneself and other people, whereby the
      perception or awareness of risk or vulnerability between two interacting
      entities is ideally symmetric.

Nomology, which concerns the study of decision-making processes and structures,
can provide the necessary framework. In nomology, a structure is a decision-
making process comprised of a series of steps or actions. One or more structures
can be combined into a framework to achieve a common purpose. Decision-making
processes largely fall into one of two generic structures. Brugha [7] argues that
decision structures can be subjective, where one is committed to a position and
seeks to persuade other people of its correctness through argument and persuasion
as in rational choice theory and game theory. Or decision structures can be
objective, where a balance between opposites is achieved through a mutual
adjustment between oneself and other people. Our definition of a social exchange
trust policy TP is based on an objective decision structure.

If such an objective approach to trust is to be successful in the intercultural
context of the mobile Internet, we need to find a conception in the East that parallels
our notions of balance or symmetry in the West. Such a conception may be found in
the moderating notion of harmony in traditional Taoist Yin-Yang culture which is
influential in countries such as China, Japan, and Korea. Du et al. [8, p. 56] have
recently proposed a moderating model of trust in conflict management based on this
notion. They argue that nomology provides a framework for combining a Taoist
Yin-Yang approach to conflict management with more conventional negotiation
and mediation strategies such as argument and persuasion. Whereas rationalistic

and deterministic approaches to conflict management attempt to reduce trust to a
calculation regarding the probability of future cooperation, a Taoist Yin-Yang
approach attempts to build trust and cooperation gradually through mutual
adjustment, moderation, and self-control. Two interacting entities can move from
cooperation to collaboration through mutual actions that increase the level of trust
or confidence in each other. This involves achieving a balance or harmony between
two sides, in contrast to one side dominating the other side as in the case of
conventional negotiation and mediation strategies.

   Although Du et al. are interested in the impact of trust on negotiation and
mediation strategies for conflict management, their Taoist Yin-Yang approach is
highly suggestive for next generation approaches to trust in open, dynamic environments such as the mobile Internet. The relevance of this approach to a virtue-
based trust model is reflected by their definition of trust as a reliance on the
integrity, ability, or character of a person with clear reference to intellectual as well
as moral virtues. Presumably, the definition also holds for social groups and systems that collectively manifest these traits of character. Asian cultures tend to
evaluate the actions of individuals and organizations rather than their beliefs, based
on traits of character such as responsibility and reputation. The relevance of this
approach to a virtue-based trust model is also reflected by the central role given in
the management of trust to adapting actions that increase balance or harmony, in
contrast to adjusting actions that reduce balance or harmony.

   Figure 6.6 shows how increasing levels of trust facilitate socially valuable ends
such as cooperation and ultimately collaboration, while decreasing levels of trust
lead to socially undesirable ends such as selfishness and ultimately conflict. The
socially undesirable end of selfishness has been added to the original diagram of Du
et al. to be consistent with Fig. 6.1. This Taoist Yin-Yang approach to conflict
management involves two patterns or forces that co-exist and interact with each
other. Whereas cooperation involves balancing the benefits and costs received by
both sides in a social exchange, selfishness involves the domination of one side by
the other side. According to our social exchange trust policy TP, the trust relation TR
is symmetric with respect to the perception or awareness of risk or vulnerability
between the interacting entities in the case of cooperation. In the case of selfishness,
however, the trust relation TR is asymmetric in that one side is subject to more risk or
vulnerability than the other side. Furthermore, whereas collaboration involves
aligning or integrating the activities of cooperating entities, conflict involves the
violation of the social exchange trust policy and the disintegration of the trust
relation based on it. Thus, as Du et al. [8, p. 57] argue, the dyadic relation between
collaboration and conflict concerns the dilemma of integration versus disintegration.

   As Fig. 6.1 shows, confrontation plays a mediating role in Fig. 6.6 between
selfishness and cooperation in the first instance, and ultimately between conflict and
collaboration. Confrontation involves smoothing or resolving issues that disrupt
balance or harmony through conflict resolution. Conflict resolution is an adaptive
process, where both entities in the social exchange mutually adjust themselves in
order to keep in balance or harmony. Whereas conflict involves taking direct

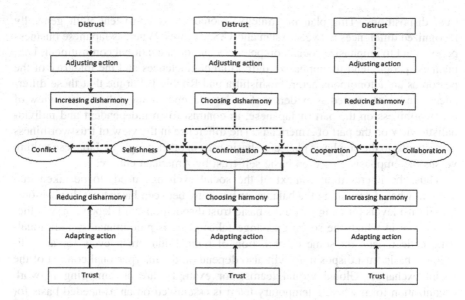

Fig. 6.6 A moderating model of trust in conflict management [8]

actions in the social exchange through the exercise of power and brute force that result in one entity dominating the other, confrontation involves taking indirect actions in the social exchange through the exercise of intellectual virtues that help mediate dyadic relations and transform them into tendency relations of the sort we want to strengthen.

In the transition from conflict and selfishness to confrontation, trust is essential to reduce imbalance or disharmony, whereby each entity no longer sees the other as an enemy and both entities feel they can co-exist. In the transition from confrontation to cooperation, each entity sees the other as a potential co-worker and both entities acknowledge they are mutually involved. In the transition from cooperation to col- laboration, each entity sees the other as a friend and both entities acknowledge they have interests and benefits in common. Progress from one stage to the next in the trust relation depends on the successful performance of mutually adapting actions by both entities. For progress to continue, Du et al. [8, p. 59] argue that balance or harmony between the entities must be established at each stage in the social interaction.

The social exchange trust policy TP is motivated by this Taoist Yin-Yang approach to conflict resolution based on increasing levels of trust through mutually adapting actions. The plan or strategy (in Foley's sense of the term) of increasing levels of trust though mutually adapting actions can be implemented by a virtue- based trust model. The trust model supports the required adapting actions through intellectual virtues that operate on and moderate the basic trust dispositions. Once it is recognized that cooperation depends on basic trust in other people and that self- ishness is the outcome of excessive basic self-trust, conflict resolution and cooper- ation can be achieved by establishing balance or harmony between these two basic trust dispositions. This plan or strategy, moreover, is consistent with generally recognized differences between Asian and U.S. cultures. Whereas Japanese business people tend to emphasize social competences and organizational commitment, U.S. business people tend to emphasize individual competences and the integrity of the person as an autonomous actor. Nishishiba and Ritchie [9] argue that these differences can be interpreted as evidence of an interdependent and collectivistic view of trustworthiness on the part of Japanese, in contrast to an independent and individualistic view on the part of Americans. The difference in the view of trustworthiness corresponds to the emphasis placed on basic trust in other people by Japanese culture versus the emphasis placed on basic self-trust by American culture.

Thus, the intercultural context of the social exchange needs to be taken into account when determining the balance or harmony between basic trust dispositions. Initial trust levels for an individual's basic trust dispositions will depend on whether the culture in which the social exchange takes place is predominantly individualistic, collectivistic, or some combination of both. Initial trust levels for an individual's basic trust dispositions will also depend on the interpersonal context of the social exchange. Global virtual teams, for example, are an emerging network organization form where a temporary team is assembled on an as-needed basis for the duration of a task and staffed by members from different countries. Bodensteiner and Stecklein [10] have done studies that show that virtual teams are more susceptible to increasing levels of distrust than traditional co-located teams due to difficulties in maintaining effective communication. Thus, one might expect that levels of distrust among members of virtual teams will increase over time, and that the appropriate corrective action would involve selecting intellectual virtues that enhance basic trust in other people while restraining excessive basic self-trust.

According to this moderating model of trust, trust is the source of adapting actions that facilitate cooperation and collaboration through the exercise of intellectual virtues. In contrast, distrust is the source of adjusting actions that lead to selfishness and conflict through the exercise of power and brute force. The motivational aspect of trust lies in its ability to create and sustain balance or harmony that leads both sides in a conflict to adjust and moderate themselves in relation to each other. As Ryutov [6, p. 91] observes, however, cooperation in social exchanges that involve the awareness of risk or vulnerability is possible only when the level of trust exceeds a minimum acceptable threshold for each side. According to computational models of trust, determining a minimum acceptable threshold is a matter of finding a set of attributes and rules that predict whether the social exchange will result in an acceptable outcome. Similarly, Coleman [11, p. 91] argues that the decision to trust other people and to cooperate with them is based on the probability that they will reciprocate. Such fixed threshold assignments for trust, however, are not adequate in open, dynamic environments such as the mobile Internet. The level of trust required in a given social exchange needs to be continuously monitored and analyzed by performing run-time evaluation of risk or vulnerability which only increases the computation and communication overhead of computational trust models.

This problem raises two questions. First, given the large computation and communication overhead of performing real-time trust computations, the question is whether we need to determine the probability of future cooperation in advance of a social exchange if there is a more efficient method for determining the level of trust required in a given context based on our definition of sufficient rationality. Secondly, given the epistemological inquiry into the nature of trust in previous chapters, the question is whether we can have or even need to have justified trust in the reliability of our sources of knowledge prior to a social exchange. In the light of these questions, there are at least two advantages to a moderating model of trust, one theoretical and the other practical. From a theoretical point of view, the model is simple in that it is based on the actions of entities observed in everyday rational activities. From a practical point of view, the model is flexible in that it can be adapted to the highly contextualized and nuanced environment of the mobile Internet. Here, as Coleman suggests, the risk one is willing to take and the vulnerability one is willing to accept depend on the performance of another actor which is consistent with a virtue-based approach to trust. Similarly, Sosa [4, pp. 188–189] argues that people need to know which members of their group are dependable. Cooperative success depends on the ability of a group to monitor the aptitudes and ineptitudes of its members in a variety of contexts and with varying objectives. The aptitudes, skills, competences, or virtues that we admire are those that are stable.

In principle, the process of evaluating whether an action is adapting or adjusting should be more efficient than the process of evaluating a belief state regarding the intentions or future actions of other entities because the evaluation does not depend on personal knowledge of the actor or entity. The action itself is morally evaluable in terms of virtue ethics without having to express the action by a belief. Although the catalog and analysis of intellectual virtues is in its infancy, one could establish criteria or a rule for the intellectual trait of open-mindedness, for example, and check to see whether a particular action satisfied the rule. The action can be considered adapting if it facilitates socially valuable ends, and the level of trust in the trust relation can be increased. Increasing levels of trust facilitate cooperation which may eventually lead to mutually beneficial forms of collaboration. In contrast to subjective trust, which is based on the belief state of one entity regarding the intentions or future actions of another entity, objective trust can be evaluated independently of belief states from data provided by an unknown entity (human or artificial) that can be objectively verified by the actions themselves. This implies that the evaluation of objective trust in the computation of trust levels could be implemented as a lightweight process, in contrast to the more demanding time and space complexity requirements of typical negotiation and mediation strategies based on rational choice or game theoretic approaches.

Since a virtue-based trust model can be implemented both as an objective decision structure and as a subjective decision structure in a nomological framework, the trust model can be seen as a complement to rather than replacement for typical negotiation and mediation strategies. An objective decision structure is consistent with our definition of responsible belief as sufficient rationality, which places practical limitations on time and intellectual or computational resources in achieving socially valuable ends such as cooperation and collaboration while avoiding socially undesirable ends such as selfishness and conflict. For the most part, trust in the social exchange can be moderated through the mutual adjustment of basic trust dispositions and the lightweight process of evaluating whether the actions taken by each side in the social exchange are adapting or adjusting. Since the evaluation involves objective trust based on an objective decision structure, personal knowledge of user identities or locations is not required, thereby protecting both source and location confidentiality.

Only when the level of trust falls below a minimum acceptable threshold, where selfish behavior may lead to conflict, do we need a subjective decision structure based on more complex and computationally expensive negotiation and mediation strategies. In confrontation, Smith [12] argues that disagreement may be seen as a source of strife that leads to conflict. Or it may be seen as an opportunity for creative interchange through mediation where new connections are made between long established and defended perspectives or boundaries. Thus, confrontation may be interpreted as an inflection point in the trust relation, where one or both sides can chose disharmony or harmony respectively. In contrast to objective trust based on an objective decision structure, typical negotiation and mediation strategies involve subjective trust based on a subjective decision structure that depends on personal knowledge of other entities and their intentions in a particular social interaction.

To be useful, however, this personal knowledge need not meet the intellectual goals of now having accurate and comprehensively coherent beliefs and full-fledged trust. According to our definition of responsible belief, which formulates a criterion for sufficient rationality, the personal knowledge need only consist of beliefs that are good enough relative to these intellectual goals given a context. This distinction is analogous to Simon's [13] distinction between substantive rationality, which involves making objectively rational choices, and procedural rationality, which involves making rational choices under resource limitations. In the case of a subjective decision structure, moreover, we need not evaluate the belief state of one entity regarding the intentions or future actions of another entity based on such personal knowledge to determine trust levels. Instead, we can evaluate acts of intellectual virtue manifested by one entity in moderating its own basic trust dispositions or reactions in response to the actions of another entity according to whether the cognitive performance achieves socially valuable ends and avoids socially undesirable ends. In the case of trust, actions speak louder than words or beliefs about promises which may be kept or broken. Thus, a virtue-based trust model is a version of trust without identity that protects confidentiality.

References

1. Zagzebski L (2014) Trust. In: Tempte K, Boyd C (eds) Vices and their virtues. Oxford University Press, New York (in press), pp 269–284. Much of Sects. 2–3 in the draft copy are based on Chap. 2 of the following book by the same author (2012) Epistemic authority: a theory of trust, authority, and autonomy in belief. Oxford University Press, New York (in

press). All references are to the draft copy: http://www.ou.edu/ouphil/faculty/zagzebski/Trust. pdf. Accessed 15 Jul 2014

2. Foley R (2012) The foundational role of epistemology in a general theory of rationality. In: Fairweather A, Zagzebski L (eds) Epistemic authority: essays on epistemic virtue and responsibility. Oxford University Press, New York, pp 214–230
3. Sosa E (2007) Apt belief and reflective knowledge. Vol. 1: A virtue epistemology. Oxford University Press, New York
4. Sosa E (2009) Apt belief and reflective knowledge. Vol. 2: Reflective knowledge. Oxford University Press, New York
5. Plato (2000) Price JT (ed) The republic (trans: Jowett B). Dover Publications, New York
6. Ryutov T (2007) A socio-cognitive approach to modeling policies in open environments. In: 8th IEEE international workshop on policies for distributed systems and networks, Bologna, Italy, 13–15 Jun 2007, pp 29–38
7. Brugha C (2006) Conflict decision processes: with illustrations from Ireland. Int Sci J Methods Models Complex 8(1):1–26. doi:10.1109/POLICY.2007.21
8. Du R et al (2011) Integrating Taoist Yin-Yang thinking with western nomology: a moderating model of trust in conflict management. Chin Manag Stud 5(1):55–67. doi:10.1108/17506141111118453
9. Nishishiba M, Ritchie LD (2000) The concept of trustworthiness: a cross-cultural comparison between Japanese and US business people. J Appl Commun Res 28(4):347–367. doi:10.1080/00909880009365581
10. Bodensteiner NM, Stecklein JM (2001) The role of a multidimensional concept of trust in the performance of global virtual teams. In: 12th international conference on comparative management, Kaohsiung, Taiwan, 23–25 May 2001, pp 1–14
11. Coleman JS (1990) Foundations of social theory. Belknap Press, Cambridge
12. Smith JE (1988) Mediation, conflict, and creative diversity. In: Liu SH, Allinson RE (eds) Harmony and strife: contemporary perspectives, east & west. Chinese University Press, Hong Kong, pp 31–48
13. Simon H (1976) From substantive to procedural rationality. In: Kastelein J et al (eds) 25 years of economic theory. Springer, Boston, pp 65–86

Chapter 7
Conclusion: Modeling Human Social Interaction

The convergence of fixed and mobile networks with other types of wireless next generation networks is rapidly evolving within a highly dynamic and unstructured environment, where the lack of a central authority forces entities to interact through collaboration and negotiation. Current computational trust models, based on the iterative exchange of personal knowledge such as digital credentials and access control policies, are not feasible due to the limited power, bandwidth, and computational capabilities of mobile devices. The computation and communication overhead of public key cryptography and policy compliance checking places excessive demands on these limited resources. In particular, He et al. [1] argue that computational trust models require pre-registration of the identities of all potential users in advance of a data exchange which is not feasible for distributed applications with strong real-time requirements. Yajun and Yulin [2] further argue that such prior knowledge is not possible in mobile ad hoc networks where nodes only temporarily collaborate. The lack of pre-authentication knowledge makes it difficult to establish initial trust between strangers.

To reduce the large computation and communication overhead of current computational trust models, Yajun and Yulin have proposed a resource-constrained trust negotiation (RCTN) procedure to build initial trust. Digital credentials and access control policies are exchanged only once at the start of the interaction, and keys are subsequently disclosed in an iterative fashion as usual. The RCTN procedure is based on a computational trust model designed for distributed environments, where trust information is obtained from peers or neighbor nodes in the network rather than from a trusted third-party. A degree or level of trustworthiness is computed for entities acting within a specific security domain based on this information. In the computational trust literature, a level of trustworthiness is a prediction about the ability of one entity to fulfill the expectations of another entity in a trust relation. Thus, trust can be interpreted as a subjective belief of one entity regarding the intentions or future actions of another entity based on its own past interactions with that entity as well as those of its neighbors.

To make current computational trust models more flexible in the highly contextualized and nuanced environment of the mobile Internet, Luca et al. [3] have proposed an adaptable, context-aware trust function to compute the level of

© The Author(s) 2014
M.G. Harvey, *Wireless Next Generation Networks*,
SpringerBriefs in Electrical and Computer Engineering,
DOI 10.1007/978-3-319-11903-8_7

trustworthiness for an agent. The trust function depends on three factors, including a presence factor that measures the experience of an agent, a frequency factor that measures how often the agent interacts with other agents, and a regularity factor that measures the persistence of an agent over time. The greater these factors, the more trustworthy the agent. Like most computational approaches to trust, however, the trust relation function depends on machine learning algorithms that have a large computation and communication overhead.

7.1 Advantages of a Virtue-Based Trust Model

These problems motivate the need for a more efficient and flexible trust model that reflects the way humans interact in social environments. Rettinger et al. [4] argue that recent studies in psychology have shown that people draw trait inferences such as trustworthiness from the mere facial appearance of strangers in a fraction of a second. That trustworthiness is typically seen as a trait of character suggests that a virtue approach to trust is well suited as a model of human social interaction. This finding also supports our conception of basic trust as a pre-reflective reaction or response. Although such initial trust lacks the quality of knowledge resulting from a person's past interactions with other people, Rettinger et al. claim that it never-theless provides a foundation for rational decision-making which is characterized more by a reflex than by a reasoning process. Their studies show that humans have the innate ability or competence to estimate the trustworthiness of other people based on contextual information which may at first glance appear unrelated to the derived expectation. In contrast, computational models of trust are based on well-defined past experiences with an entity, from which it is allegedly possible to predict that the entity will act in the future as it did in the past regardless of the context of the interaction. This assumption has proven problematic in the highly contextualized and nuanced environment of the mobile Internet.

Perhaps the single most important advantage of a virtue-based trust model, in contrast to trust models based on identity, roles or attributes, is that it reflects both the instinctual rational conduct and reflective rational conduct of humans. Most of the time, human social interactions operate on automatic pilot. Humans instinctively react to each other in a trustful way at the level of animal knowledge until there is a violation of basic trust that requires negotiation and mediation at the level of reflective knowledge. For this reason, we have called basic trust an animal competence or ability that is not reason-based, whether this trust is directed toward one's own self or toward other people. Basic trust, in both of its forms, is part of our natural endowment as well-functioning epistemic beings. At the level of animal knowledge, Wittgenstein [5, Sect. 287] suggests that we do not consciously predict the future actions or behavior of other people any more than the squirrel predicts that it is going to need stores for next winter as well. At the level of reflective knowledge, however, we have to justify our initial trust in other people with better knowledge through the exercise of reason-based competences such as intellectual virtues which help us avoid being either too trustful of other people or too suspicious of them.

We have seen how such individual competences can be generalized to social competences, and how social competences can be manifested by systems as well as by human groups. This abstraction allows us to extend the distinction between animal knowledge and reflective knowledge in individual epistemology to autonomous rational agents acting in a social context. Gmytrasiewicz and Durfee [6] argue that rational agents that are fully autonomous should be able to freely choose among all actions available to them, including physical and communicative actions. The challenge in designing such a system is to develop the theories and methods needed by the autonomous systems to use their capabilities rationally. A virtue-based trust model may be seen as the first step toward the development of such a theory, though its methods need to be defined. Physical actions can be interpreted as either adapting actions that increase levels of trust, cooperation and collaboration, or as adjusting actions that increase levels of distrust, selfishness and conflict. Communicative actions include both types of physical actions, which can be expressed in the form of beliefs that can be evaluated by a mediation process, if necessary. Instead of trying to change beliefs through argument and persuasion, however, the basic trust dispositions that motivate and influence the beliefs can be moderated through the exercise of appropriate intellectual virtues. Intellectual virtues can be interpreted as reason-based competences that help us do a better job of what we do naturally when our ground-level, animal competences fail us.

A virtue-based trust model has two additional advantages over strictly cognitive and computational approaches to trust, one theoretical and the other pragmatic. The theoretical advantage of the trust model lies in its definition of sufficient or practical rationality in terms of a general theory of rationality anchored in the more restrictive and rigorous notion of epistemic rationality or normativity. The distinction between sufficient rationality (responsible belief) and epistemic rationality (epistemically rational belief) is analogous to Simon's [7] distinction between procedural rationality and substantive rationality respectively. This general theory of rationality allows us to relate trust and rationality in a non-circular fashion through the interaction and cross-level coherence of beliefs at the animal and reflective levels. The theory not only allows us to interpret basic trust as the necessary condition and foundation of rationality. It allows us to interpret trust and reason as complementary epistemic mechanisms or competences that guide our rational conduct at two different epistemic levels which interact at a more reflective level in the presence of the awareness of risk or vulnerability.

The pragmatic advantage of the trust model lies in its central notion of virtue which plays a key role in facilitating rational activities in a global context that depend on cooperation and collaboration. The trust model provides norms or criteria based on an examination of human knowledge that allow us to evaluate the epistemic success or reliability of the performance of a human or artificial rational agent, measured by how well the performance achieves socially valuable ends such as cooperation and collaboration and avoids socially undesirable ends such as selfishness and conflict. The central role played by virtue and character across cultures, moreover, suggests that a virtue-based trust model can be adapted to the intercultural context of the mobile Internet. As such, a virtue-based trust model not only provides a normative basis for evaluating current trust models, but a theoretical basis for developing more efficient and flexible trust models in the future.

7.2 General Features of a Virtue-Based Trust Model

A virtue approach to trust is based on the conception of a social exchange trust policy, where two interacting entities enter into a trust relation. The trust relation is based on the willingness of both entities to accept more or less the same degree of risk or vulnerability in the social interaction. Increasing levels of trust and cooperation are manifested by mutually adapting actions that keep the trust relation in balance or harmony. In contrast, increasing levels of distrust and selfishness are manifested by adjusting actions, whereby one entity attempts to dominate the other entity in the trust relation. The goal of the social exchange trust policy is to keep the trust relation symmetric. Thus, when one entity dominates another entity through selfish behavior by accepting far less risk or vulnerability in the trust relation, the level of trust should be decreased for the selfish entity. Conversely, when one entity moderates its own reactions in response to the actions of another entity, the level of trust should be increased for the cooperative entity.

The trust model is based on a general theory of rationality that allows us to integrate the cognitive and behavioral aspects of trust in a nomological framework. For most social interactions, the trust relation can be evaluated in an efficient fashion according to an objective decision structure using simple rules based on actions that are either adapting or adjusting. In this case, the trust model reflects how we do what we do naturally at the level of animal knowledge until there is a violation of basic trust. Only when the level of trust falls below a minimum acceptable threshold, where selfish behavior may lead to conflict, is a subjective decision structure needed based on more computationally expensive negotiation and mediation strategies. In this case, the trust model reflects how we can do a better job of what we do naturally by justifying our initial trust in the reliability of entities or sources of knowledge at the level of reflective knowledge through the exercise of intellectual virtues. The goal of justified trust involves the intellectual goals of full-fledged trust and epistemically rational belief, both of which can be limited in accordance with the security domain and security level required for a specific social interaction.

One of the insights of a virtue-based trust model is the mediating role played by beliefs between actions and entities. Actions or reactions on the part of one entity toward another entity, whether trustful or distrustful, may be expressed in the form of explicit beliefs that are epistemically evaluable. Beliefs, regardless of how entrenched they may be in our cognitive structure, can be changed by the exercise of appropriate intellectual virtues. To avoid the large computation and communication overhead associated with the evaluation of belief states regarding the intentions or future actions of other entities, however, we need not evaluate the beliefs themselves based on personal knowledge of the identities of other entities. Instead, we can evaluate acts of intellectual virtue manifested by one entity in moderating its own basic trust dispositions or reactions in response to the actions of another entity according to whether the cognitive performance achieves socially valuable ends and avoids socially undesirable ends. In the case of trust, actions speak louder than words or beliefs about promises which may be kept or broken. Thus, a virtue-based trust model is a version of trust without identity that protects confidentiality.

According to a virtue-based trust model, initial basic trust is established instinctually on the basis of innate dispositions or competences absent prior knowledge of the past interactions of an entity with other entities. The innate disposition or reaction to trust other people corresponds to the start state of basic trust in our virtue-based trust model. This initial epistemic state is consistent with the empirical findings of McKnight et al. [8], who have developed an initial trust model based on attribution theory to explain the presence of high initial trust-worthiness in newly formed relationships such as temporary virtual teams. Like their initial trust model, a virtue-based trust model helps us see that trustworthiness initially has the epistemic status of an instinctive reaction or implicit commitment which may or may not be expressed by an explicit belief, and that this implicit commitment precedes rational trust or full-fledged trust. In contrast to computa-tional models of trust, McKnight et al. argue that initial, pre-reflective trust is not based on conscious inferences about the likelihood of future cooperation of other people in assisting us to achieve our goals. Instead, like our virtue-based trust model, people rely on their own pre-existing dispositions, cognitive processes, or epistemic competences to make attributions about the initial trustworthiness of other people at a more instinctual level which are preponderantly true, even if occasionally false.

Although the trust relation begins in an epistemic state of basic trust, the trust relation needs to develop toward the goal state of full-fledged trust (justified trust) as the awareness of risk or vulnerability in the trust relation increases. According to attribution theory, the causes of other people's actions are attributed to internal characteristics or properties of the person so long as his or her behavior is consistent with prior expectations. When the behavior of other people is inconsistent with prior expectations, however, the causes of their actions are attributed to external situational characteristics. McKnight et al. argue that trust development is an attributional process, where social perceptions and beliefs are formed about other people as we try to explain their past actions. Similarly, according to a virtue-based trust model, people trust other people so long as there is no violation of basic trust. When a trust violation occurs, one entity needs to justify its initial trust in another entity or source of knowledge by establishing the reliability of the source. Here the awareness of risk or vulnerability in the trust relation plays an analogous role to skepticism in epistemology. Just as the awareness that our knowledge is vulnerable to mistakes motivates us to do a better job reflectively of what we do naturally, so the awareness that sensitive information is vulnerable to threats motivates us to do a better job of protecting this information. Thus, a well-defended system is analogous to a well-defended epistemic perspective in our trust model.

The establishment of initial trust levels in the trust relation needs to take into account contextual factors that can influence the innate dispositions or natural reactions to trust oneself and other people. First, initial trust levels for a particular social interaction depend on the intercultural context of the interaction. Fulmer and Gelfand [9], for example, have examined how cultural differences in self-understanding can influence the relation between the magnitude of a trust violation and changes in the level of trust in trust dissolution and trust recovery. In our virtue-based trust model, trust dissolution corresponds to increasing levels of distrust and disharmony in the trust relation, whereas trust recovery corresponds to increasing levels of trust and harmony. In both cases, changing trust levels are predominantly influenced by whether the entity is a member of a class or culture that sees itself as individualistic or as collectivistic. Thus, whereas one entity in the trust relation may be more individualistic where basic self-trust or autonomy is the dominant factor, the other entity in the trust relation may be more collectivistic where basic trust in other people or organizational commitment is the dominant factor. Hofstede et al. [10] argue that this is one of the four dimensions of culture. Given the pervasive influence of this dimension of culture on intercultural interactions, they argue that rules need to be formulated for individualistic versus collectivistic agent behavior.

Secondly, initial trust levels for a particular social interaction also depend on the interpersonal context of the interaction. In particular, as global virtual teams become more important in facilitating collaboration in science, politics and economics, the trust relation between team members needs to take into account the empirical finding that virtual teams are more susceptible than traditional co-located teams to increasing levels of distrust due to the potential for ineffective communication.

7.3 Summary

The convergence of fixed and mobile networks with other types of wireless next generation networks in the ongoing evolution of the mobile Internet will create several technical challenges and privacy and security issues that are not adequately addressed by current trust models.

- Current computational trust models cannot support the decentralized control and self-organizing behavior of wireless next generation networks. Secure communication in an open and dynamic environment such as the mobile Internet cannot utilize centralized authentication and authorization services due to the limited power, bandwidth, and computational capabilities of mobile devices.
- Current computational approaches to trust, based on the iterative disclosure of personal knowledge such as digital credentials and access control policies, are

not feasible in open, dynamic environments due to the large computation and communication overhead of these processes. Personal knowledge of other entities required for pre-authentication is not feasible for mobile ad hoc networks whose nodes only temporarily collaborate, or for distributed applications with strong real-time requirements.

- The task of maintaining identities for all potential users is not feasible due to the rapid growth of mobile devices and the continuously changing profile of user populations. The lack of prior personal knowledge about other entities makes the establishment of initial trust difficult in wireless next generation networks.
- Since wireless next generation networks lack a universal trust model, an entity wishing to gain access to n different Internet-based services needs to potentially undergo n distinct registration processes. Thus, personal information could be stored in n different databases which compromises confidentiality. This problem motivates the need for local information sharing schemes.
- The calculation of trust levels and the prediction of future actions based on the interaction history of an entity in computational trust models involves large computation and communication overhead which compromises availability. Since the trustworthiness of entities can vary over time, continuous updating of trust models is required which only adds to the computation and communication overhead.
- Next generation approaches to trust need to address secure handovers as mobile devices are enabled to roam between networks with different access technologies, security policies, security domains, and security levels. These handovers need to be performed efficiently while roaming between different networks to prevent leakage of secret keys and network service disruptions which compromise confidentiality and availability respectively.
- Next generation approaches to network security need to address secure routing, data aggregation, and storage. These approaches should be based on novel local information sharing schemes deployed over local areas using wireless sensor networks and real-time distributed control systems.

These technical challenges and privacy and security issues make it clear that current trust models, whether based on identity, roles or attributes, have proven neither useful nor efficient for establishing trust between strangers in open, dynamic environments such as the mobile Internet. A more efficient and flexible trust model is needed that reflects human social interaction and supports the ad hoc creation of social exchange trust policies for interactions within specific security domains and intercultural and interpersonal contexts.

- Recent sociological approaches to trust place a strong emphasis on mutual adjustment, moderation, and self-control which is consistent with a virtue approach to trust. These approaches are increasingly supported by computer scientists, who argue that next generation approaches to trust need to model the way people interact in social environments. Sociologists further argue that these approaches need to explain the relation between trust and rationality in a non-circular fashion.

- Rational trust, or being properly trusting, can be defined as the mean between two irrational extremes–over-confidence on one hand, where one entity is too trusting of another entity, and over-diffidence on the other, where one entity is not trusting enough of another entity. This definition of rational trust motivates the conception of the trust relation as the mutual adjustment, moderation, and self-control of two basic trust dispositions, self-trust and trust in other people, through the exercise of an appropriate mix of intellectual virtues.
- The trust relation is a form of social exchange, where the distribution of risk and the acceptance of vulnerability should be more or less symmetric between two interacting entities. The balance or harmony between the entities in the trust relation can be continuously monitored in a basic trust feedback loop over a local area using wireless sensor networks and real-time distributed control systems.
- Initial trust between strangers can be established on the basis of properties of persons at the level of animal knowledge, in contrast to prior knowledge of persons at the level of reflective knowledge. These properties include the basic trust dispositions which are innate competences that are rational but not reason-based. Since the deliverances of these dispositions are preponderantly true, even if occasionally false, they are reliable mechanisms with a minimal sort of justification that can be used to establish initial trust.
- The basic trust dispositions can be enhanced or restrained by intellectual virtues at the level of reflective knowledge relative to the actions of other entities. The level of trust in a given social interaction can be determined by evaluating acts of intellectual virtue manifested by one entity in moderating its own basic trust dispositions or reactions in response to the actions of another entity, in contrast to evaluating the belief state of one entity regarding the intentions or future actions of another entity based on personal knowledge of the entity.
- The process of evaluating actions rather than belief states should be more computationally efficient, since simple rules can be formulated for distinguishing adapting actions from adjusting actions. Whereas adapting actions increase the level of trust and facilitate cooperation and collaboration, adjusting actions increase the level of distrust and lead to selfish behavior and conflict. More complex rules can be formulated for determining whether an entity has manifested a certain intellectual virtue in the moderation of its basic trust dispositions.
- Intellectual virtues are reason-based dispositions or competences that enable us to do a better job of what we do naturally after there is a violation of basic trust. Intellectual virtues are required at the level of reflective knowledge to justify initial trust in other entities by monitoring, controlling, and moderating the relative influences of the basic trust dispositions operative at the level of animal knowledge. To increase the level of trust and facilitate cooperation in the trust relation, an entity needs to avoid selfish behavior that could lead to conflict by practicing intellectual virtues that enhance basic trust in other entities while restraining excessive basic self-trust.

- To avoid being too trusting of other entities in the trust relation, an entity needs to select intellectual virtues that enhance basic self-trust while restraining excessive basic trust in other entities. The goal of the trust relation is to select an appropriate mix of complementary intellectual virtues that depend on the security domain and security level of a given social interaction which will achieve balance or harmony in the trust relation.
- A virtue-based trust model can be used to support local information sharing schemes such as secure key management over localized areas using wireless sensor networks and real-time distributed control systems. Mobility can be leveraged in such schemes to spatially diffuse secret information to be protected throughout the monitored area in an energy efficient fashion. Local information sharing schemes can reduce the bandwidth and energy consumption requirements of mobile devices within acceptable constraints. Most importantly, they can provide availability without leakage of secret keys and loss of confidentiality.
- A virtue-based trust model is a version of trust without identity that protects confidentiality, since it is based on the actions of other entities rather than on personal knowledge of the actor. If used to support local information sharing schemes, a virtue-based trust model can also help ensure availability. In addition, a virtue-based trust model can be adapted to the intercultural context of the mobile Internet, given its emphasis on virtue and character as universal traits of trustworthiness and its moderating notion of achieving balance or harmony in the trust relation.

Future research needs to address several matters. Since a virtue-based trust model is motivated by the intuition that it should be more efficient to evaluate the actions, competence, or performance of an agent than it is to evaluate the belief state of one agent regarding the intentions or future actions of another agent, new methods need to be defined. Simple rules need to be formulated for distinguishing adapting actions from adjusting actions in accordance with their social ends. In the case of confrontation, where negotiation and mediation are unavoidable, more complex rules need to be formulated for determining whether an entity has manifested a certain intellectual virtue or competence that is conducive to increasing the level of trust in a given social interaction. This requires a catalog of intellectual virtues and an analysis of how each virtue affects and moderates the basic trust dispositions, both of which are currently lacking in the virtue epistemology literature.

A virtue-based trust model also needs to be defined for various attack vectors for wireless next generation networks. It remains to be made clear how the trust model can be implemented in different areas of network security such as secure routing, where an entity must trust a node to correctly route a packet through the network, or secure key management, where an entity in the serving network must trust another entity in the same network to derive and distribute keys to a target network.

Finally, in order to be robust, a virtue-based trust model needs to be able to respond to atypical interaction scenarios where intellectual virtues are intentionally misused as in the case of hacking, or where the distinction between intellectual virtues and vices is blurred as in the case of ethical hacking. Although the intention of the first action is malicious while the intention of the second action is benevolent, both actions require the exercise of similar intellectual virtues such as carefulness, thoroughness, and open-mindedness in the sense of being able to think outside the box. Thus, it is not clear how intellectual virtues can be used to distinguish between these atypical interaction scenarios without evaluating the belief state of one entity regarding the intentions or future actions of another entity and the specific security domain in which they are expressed.

References

1. He Y et al (2008) An efficient and minimum sensitivity cost negotiation strategy in automated trust negotiation. In: 2008 international conference on computer science and software engineering, Wuhan, China, 12–14 Dec 2008, pp 182–185
2. Yajun G, Yulin W (2007) Establishing trust relationship in mobile ad-hoc network. In: 3rd international conference on wireless communications, networking, and mobile computing, Shanghai, China, 8–10 Oct 2007, pp 1562–1564
3. Luca L et al (2009) Enabling adaptation in trust computations. In: Computation world: future computing, service computation, cognitive, adaptive, content, patterns (computationworld '09), Athens, Greece, 15–20 Nov 2009, pp 701–706
4. Rettinger A et al (2007) Learning initial trust among interacting agents. In: Cooperative information agents XI: 11th international workshop (CIA 2007), Delft, The Netherlands, 19–21 Sept 2007, pp 313–327
5. Wittgenstein L (1969) Anscombe GEM, von Wright GH (eds) On certainty (trans: Paul D, Anscombe GEM). Harper Torchbooks, New York
6. Gmytrasiewicz PJ, Durfee EH (1993) Toward a theory of honesty and trust among communicating autonomous agents. Group Decis Negot 2(3):237–258. doi:10.1007/ BF01384248
7. Simon H (1976) From substantive to procedural rationality. In: Kastelein J et al (eds) 25 years of economic theory. Springer, Boston, pp 65–86
8. McKnight DH et al (1998) Initial trust formation in new organizational relationships. Acad Manag Rev 23(3):473–490
9. Fulmer CA, Gelfand MJ (2009) Are all trust violations the same? A dynamical examination of culture, trust dissolution, and trust recovery. In: Modeling intercultural collaboration and negotiation (MICON) workshop, Pasadena, CA, 13 Jul 2009, pp 56–65
10. Hofstede GJ et al (2008) Individualism and collectivism in trade agents. In: New frontiers in applied artificial intelligence: 21st international conference on industrial, engineering and other applications of applied intelligent systems, Wrocław, Poland, 18–20 Jun 2008, pp 492–501

Glossary

Abduction An inference to the best explanation that assigns special status to explanatory considerations. Abduction is one of three types of inference, the other two being deduction and induction. In deductive inferences, what is inferred is necessarily true if the premises from which it is inferred are true. In contrast, inductive inferences are strictly based on statistical data such as observed frequencies of occurrences of a particular feature in a given population.

Access control The process of mediating a request for resources and data maintained by a system, and determining whether the request should be granted or denied. The access control decision is enforced by a mechanism that implements regulations established by a security policy. A security policy defines the high-level rules according to which access control must be regulated.

Amplification A distributed denial-of-service (DoS) attack intended to increase the scope, magnitude, and impact of DoS attacks.

Automated Trust Negotiation (ATN) An approach to regulate the exchange of personal knowledge such as digital credentials and access control policies during the process of establishing mutual trust between strangers wishing to share resources or conduct business transactions. Digital credentials and access control policies are treated as sensitive resources whose access is regulated by access control policies.

Autonomous Rational Agent An agent is anything that perceives its environment through sensors and acts upon that environment through effectors. A rational agent is one that performs an action that will cause the agent to be most successful. A rational agent is autonomous to the extent that its actions or behavior are determined by its own experience rather than by built-in knowledge used to construct the agent for a particular environment. However, just as evolution provides animals with enough built-in reflexes to enable them to survive long enough to learn for themselves, an autonomous rational agent must have some initial knowledge as well as the ability to learn. A fully autonomous rational agent should be able to operate successfully in a wide variety of environments or contexts such as the mobile Internet.

© The Author(s) 2014
M.G. Harvey, *Wireless Next Generation Networks*,
SpringerBriefs in Electrical and Computer Engineering,
DOI 10.1007/978-3-319-11903-8

Authentication The process of verifying the identity of an entity.

Authorization The process of determining what is permitted for an authenticated entity.

Availability The property of being able to access and use information systems in a timely and reliable fashion by authorized users. A loss of availability involves the disruption of access to or use of information systems, or denial of service. The more critical a component or service, the higher the level of availability is required.

Code Division Multiple Access (CDMA) A digital technique that provides communication services to multiple users in a single bandwidth medium. This technique is also called direct sequence spread spectrum (DSSS) because it takes the digitized version of an analog signal and spreads it over a wider bandwidth at a lower power level. This allows the digital voice to be shared with other users using different codes.

Cognitio versus *Scientia* A distinction introduced by the philosopher René Descartes to indicate the qualitative difference between two varieties of knowledge and certainty. Whereas *cognitio* involves a clear awareness of an object or proposition, *scientia* involves true knowledge of the object or proposition. *Scientia* is thought to be a higher and more certain form of knowledge compared to *cognitio*.

Coherentism A theory of knowledge according to which a belief is justified or justifiedly held if the belief coheres with a set of beliefs, and the set of beliefs forms a coherent system.

Confidentiality The property of not being disclosed to unauthorized entities. More broadly, confidentiality refers to the concealment of information or resources. Access control mechanisms support confidentiality. One such access control method is cryptography which scrambles confidential data to render it incomprehensible. A cryptographic key controls access to the unscrambled data, but the key itself becomes another datum to be protected by a cryptographic key management system (CKMS).

Cryptographic Key Management System (CKMS) A cryptographic key is a parameter used with a cryptographic algorithm that determines its operation. An entity with knowledge of the key can reproduce or reverse the operation, but an entity without knowledge of the key cannot. A key management system is responsible for handling cryptographic keys and related security parameters during the entire lifecycle of the keys. Key management is supported by a key management infrastructure that includes the framework and services needed for the generation, distribution, control, accounting, and destruction of all cryptographic material, including symmetric (private) keys, public keys, and public key certificates.

Defense-in-Depth A security strategy in which more than one security control or class of controls is used to provide overlapping or layered protection. Controls can be grouped into three largely independent classes. Physical controls prevent an attack by using walls, fences, locks, human guards, or sprinklers. Procedural or administrative controls require or advise people how to act such as laws and regulations, policies, procedures, and guidelines. Technical controls counter threats with hardware or software technology such as passwords and access controls enforced by an operating system or software application.

Denial-of-Service (DoS) An attack that generates several session initiation requests to a network host, thereby creating network traffic congestion that prevents authorized users from gaining access to a network.

Encryption The process of encoding a message so that its meaning is not obvious. A cryptosystem involves a set of rules for how to encrypt the plaintext and decrypt the ciphertext. The encryption and decryption rules are encoded by algorithms, and often use a parameter called a cryptographic key. The resulting ciphertext depends on the original plaintext message, the algorithm used to encrypt the message, and the value of the key.

Eavesdropping An attack that involves obtaining access to sensitive information such as routing addresses and private user identities without the knowledge or consent of the communicating entities.

Epistemology (Theory of Knowledge) One of the core areas of philosophy which is concerned with the nature, sources, and limits of knowledge. Although there is a vast array of theories of knowledge, most epistemologists agree that knowledge involves more than mere true belief. For example, lucky guesses or true beliefs resulting from wishful thinking are not knowledge. Thus, a central question in epistemology is what must be added to true beliefs to convert them into knowledge.

Externalism versus Internalism Internalism claims that a person either does or can have access to the basis for knowledge or justified belief, and that the person either is or can be aware of this basis. In contrast, externalism denies that one can *always* have this sort of access to the basis for one's knowledge and justified belief. Another form of internalism concerns the basis for rather than access to a justified belief. Mentalism, for example, claims that a belief is justified by a mental state of the epistemic agent holding the belief. In contrast, externalism claims that something other than mental states operate as justifiers.

Fixed/Mobile Convergence (FMC) An emerging technology that aims to create a unified communications infrastructure composed of fixed and wireless mobile networks. This converged communications infrastructure will allow users to roam seamlessly across networks and access services using different devices.

Foundationalism A theory of knowledge according to which all knowledge, or justified belief, rests ultimately on a foundation of non-inferential knowledge or justified belief. Foundationalism contrasts inferential knowledge with a kind of knowledge that does not involve having previous knowledge. Although classical foundationalism was taken to be trivially true for thousands of years, it has become the subject of intense debate in contemporary epistemology.

Frequency Division Multiple Access (FDMA) A digital technique that provides communication services to multiple users by dividing one channel or bandwidth into multiple individual bands or subchannels, each of which can be allocated to a different user. Each individual band or subchannel is wide enough to accommodate the signal spectra of the transmissions to be propagated. The data to be transmitted is modulated on to each subcarrier, and all of them are linearly mixed together. The best example of FDMA is the cable television system in which a single coax cable is used as the medium to broadcast hundreds of channels of video/audio programming to homes.

Handover The process of transferring an active call or data session from one cell in a cellular network to another, or from one channel in a cell to another. An efficient handover scheme is important for delivering uninterrupted service to a caller or data session user. Handovers can be classified into two types, hard or soft, the first being less costly to implement than the second. Whereas a hard handover is characterized by an actual break in the connection while switching from one cell or base station to another, a soft handover involves two connections to the cell phone from two different base stations to ensure that no break occurs during the handover.

Hop A direct communications channel between two computer systems. In a complex computer network, a message might take several hops between its source node and its destination node.

Injection Attack An attack that provides some form of input, but attaches additional malicious data to perform a command or include additional input. Injection attacks typically occur when input has not been validated.

Integrity The property of not being able to modify confidential data except by authorized entities. More broadly, integrity refers to the trustworthiness of data or resources. Whereas data integrity refers to the trustworthiness of the content of the information, origin integrity refers to the trustworthiness of the source of the information. The source of the information may bear on its accuracy and credibility, and can thus affect the level of trust that people place in the information. A digital signature is a security mechanism that provides both data integrity and origin integrity.

Jamming Attack A denial-of-service (DoS) attack of radio interference whose goal is to disrupt communications between authorized entities. In wired networks, DoS attacks are typically performed by filling user domain and kernel domain buffers. In wireless networks, an attacker can launch more severe DoS attacks by continuously emitting RF signals to fill a wireless channel, thereby blocking the wireless medium and preventing authorized entities from communicating.

Kp Requirement for Apt Belief For any correct belief (or presupposition) that p, its correctness is attributable to a competence only if it derives from the exercise of that competence in conditions appropriate for its exercise, where that exercise in those conditions would not too easily have issued a false belief (or presupposition). See Chap. 3.

KR Requirement of Internalism A potential knowledge source K can yield knowledge for subject S only if S knows that K is reliable. See Chap. 3.

Man-in-the-Middle Attack An attack that uses a network node as the intermediary between two other nodes. Each of the endpoint nodes believes that it is communicating or interacting with the other, but each is actually interacting with the intermediary.

Masquerading Attack An attack that involves pretending to be some entity without the permission of that entity, typically to gain unauthorized access to information.

Message Body Tampering An attack that involves altering the content of a message, thereby compromising its data integrity.

Mobile Internet The ongoing evolution of mobile networks toward an architecture encompassing an IP-based core network and different wireless access networks. Signaling with the core network is based on IP protocols that are independent of the access network. This allows the same IP-based services to be accessible over any networked system. The IP multimedia core network subsystem (IMS), standardized by the third-generation partnership project (3GPP) and third-generation partnership project 2 (3GPP2), is the first commercial approach toward an IP-based core network.

MR Requirement of Internalism In order to understand one's knowledge satisfactorily, one must see oneself as having some reason to accept a theory that one can recognize would explain one's knowledge if it were true. See Chap. 4.

Next Generation Network (NGN) A packet network that provides telecommunications services to users, and utilizes multiple broadband and quality of service (QoS)-enabled transport technologies. The service-related functions are independent of the underlying transport-related technologies. NGNs provide unfettered access to networks, services and service providers, and support mobility that will enable consistent and ubiquitous provisioning of services to users.

Nonrepudiation The property of being able to verify that an operation has been performed by a particular entity or account. This is a system property that prevents an entity in an online interaction from subsequently denying involvement in the interaction.

Orthogonal Frequency Division Multiple Access (OFDMA) A digital technique used in long-term evolution (LTE) cellular systems to accommodate multiple users in a given bandwidth. The modulation technique divides a channel into multiple narrow orthogonal bands that are spaced so that they do not interfere with each other. The data to be transmitted are divided into many lower-speed bit streams, modulated onto subcarriers, and assembled into time segments that are transmitted over the assigned subcarriers. A group of subcarriers is assigned to each user which includes subchannels and related time slots.

PC Principle of the Criterion Knowledge is enhanced through justified trust in the reliability of its sources. See Chap. 3.

PC* Principle of the Criterion Knowledge is enhanced through justified trust in the reliability of its sources, which depend on natural reactions that lie beyond being justified or unjustified. See Chap. 4.

Pre-authentication The process of verifying that an entity knows a password before it is allowed access to a high quality secret that has been encrypted with the password. Pre-authentication prevents attackers from easily obtaining a quantity that could aid off-line password guessing to gain unauthorized access to information.

Privacy The property of being protected from the unauthorized disclosure of personal information. In computer security, the term confidentiality is often used instead because the term privacy has been co-opted by the legal profession to refer to the opposite. Privacy legislation consists of laws that require governments and businesses to inform people about what personal information is being collected about them.

Public Key Infrastructure (PKI) An infrastructure for binding a public key to a known entity through a trusted intermediary such as a certificate authority. Public key cryptography, also called asymmetric cryptography, is a cryptographic system where encryption and decryption are performed using different keys, a private key that must be kept secret and a public key that can be safely divulged respectively. In contrast, secret key cryptography, also called symmetric cryptography, is a cryptographic system where the same secret key is used both for encryption and decryption.

Real-Time Distributed Control System (RTDCS) A system that integrates computing and communication capabilities with monitoring and control of entities in the physical world. Systems are typically composed of a set of networked agents, including sensors, actuators, control processing units, and communication devices. Although some forms of RTDCSs are already in use in areas such as supervisory control and data acquisition (SCADA) systems, the widespread growth of wireless embedded sensors and actuators is creating new applications in areas such as medical devices, autonomous vehicles, and smart structures.

Registration Hijacking An attack that involves the use of stolen personal information for sending a bogus registration request to gain access to a network.

Reliabilism A theory of knowledge that emphasizes the truth-conduciveness of a belief-forming process, method, or other epistemically relevant factor.

Replay Attack An attack that involves storing and retransmitting messages with the aim of disrupting service in some region of a network.

Risk The possibility of suffering a loss. Risks depend on the enterprise context or environment and may change with time. Although many risks are quite unlikely to occur, they still exist and should be addressed in a thorough risk assessment.

Risk Assessment The process of analyzing an environment to identify the threats, vulnerabilities, and mitigating actions in order to determine the impact of an event affecting a project, program, or enterprise.

Risk Management The overall decision-making process of identifying threats and vulnerabilities and their potential impacts to an environment, determining the cost of mitigating such events, and deciding what actions are most cost effective in controlling these events.

Roaming A wireless network service extension in an area that differs from the registered home network location. Roaming allows a mobile device to access the Internet and other mobile services when out of its normal coverage area. It also gives a mobile device the ability to move from one access point to another. Roaming services are usually provided by cellular service providers and Internet service providers (ISPs) through a cooperative agreement.

Routing Attack An attack that involves interference with the correct routing of packets through a network. Several different types of routing attacks can be carried out at the network layer, including spoofed, altered, or replayed routing information. These attacks can create routing loops, extend or shorten intended routing paths, generate bogus error messages, and increase end-to-end latency, thereby compromising availability.

S (General Schema for Rationality) A plan (decision, action, strategy, belief, etc.) is rational in sense X for an individual if it is epistemically rational for the individual to believe that it would acceptably contribute to his or her goals of type X. See Chap. 5.

Security Association (SA) A shared state between two entities in an online interaction such as a cryptographic key, the identity of the other entity, or a sequence number, as well as the cryptographic algorithm to be used for encryption and decryption.

Security Silo The traditional process of securing individual components of an information technology infrastructure such as networks, desktops, database systems, or software applications. A broader approach is to integrate security technologies across these various components to ensure that their interactions are also secure.

Selective Forwarding Attack An attack that subverts a node in a network in order to drop selected packets.

Server Impersonation An attack that involves the redirection of network traffic and routing requests to a malicious proxy server.

Service Level Agreement (SLA) A contract between a service provider and a subscriber that details the nature, quality, and scope of the service to be provided.

Service-Oriented Architecture (SOA) The latest evolution of distributed architectures based on the client-server model. SOA implementations are centered around the idea of a service which refers to a modular, self-contained piece of software that has well defined functionality. The service is expressed in abstract terms independently of the underlying implementation. All SOA implementations include a service provider, a service requestor, and a service registry. Service providers can publish the details about a service to a service registry, where service requestors can obtain the details. The service requestor can then invoke (bind) the service on the service provider. Web services is currently the most popular form of SOA implementation.

Session Tear-Down An attack that disconnects authorized users from a network, requiring them to reconnect several times, thereby creating network traffic congestion that disrupts network service.

Single Carrier Frequency Division Multiple Access (SC-FDMA) A digital technique that assigns multiple users to a shared communications resource. SC-FDMA can be interpreted as a linearly precoded orthogonal FDMA (OFDMA) scheme, since it has an additional discrete Fourier transform (DFT) processing step that precedes the conventional OFDMA processing (see OFDMA). The technique has become an attractive alternative to OFDMA in uplink communications where lower peak-to-average power ratio (PAPR) greatly benefits the mobile terminal in terms of transmit power efficiency and reduced cost of the power amplifier. For this reason, it has been adopted as the uplink multiple access scheme in 3GPP long-term evolution (LTE), namely, evolved UTRA (E-UTRA).

Sinkhole Attack An attack that subverts a node in a network in order to attract packets to it.

Third Generation (3G) Cellular Network A mobile telecommunications technology based on a set of standards for mobile devices and mobile telecommunications services and networks. The set of standards comply with the International Mobile Telecommunications-2000 (IMT-2000) specifications developed by the International Telecommunication Union (ITU). Applications of 3G include wireless voice telephony, mobile Internet access, fixed wireless Internet access, video calls, and mobile TV.

Threat A potential breach of confidentiality, integrity, or availability. In the case of confidentiality, the privacy of an entity is potentially violated. In the case of integrity, data is potentially changed or comes from an untrusted source. In the case of availability, an asset or resource potentially becomes unavailable due to a denial-of-service (DoS) attack. Threats can come from outside or inside some boundary or perimeter that defines the system. They can come from authorized entities or unauthorized entities masquerading as authorized entities in order to bypass security mechanisms. Threats can also come from human errors or environmental disruptions.

Time Division Multiple Access (TDMA) A digital technique that divides a single channel or band into time slots. Each time slot is used to transmit one byte, or another digital segment of each signal, in sequential serial data format. The technique works well both for slow voice data signals and for compressed video and other high-speed data. TDMA is widely used for T1 transmission systems to carry up to 24 individual voice telephone calls on a single line. The global system for mobile communications (GSM) cellular phone system is also TDMA-based. GSM divides the radio spectrum into 200-kHz bands and uses time division techniques to place eight voice calls into one channel.

TP (Social Exchange Trust Policy) A social exchange trust policy aims to achieve a balance between opposites through a mutual adjustment between oneself and other people, whereby the perception or awareness of risk or vulnerability between two interacting entities is ideally symmetric. See Chap. 6.

Traffic Analysis An attack that involves deducing properties of a data exchange based on personal knowledge of the interacting entities, the duration of the data exchange, timing, bandwidth, and other technical characteristics that are difficult to disguise in packet networks. Traffic analysis is used in conjunction with eavesdropping in passive attacks to monitor network traffic and wireless communication channels for information that can be used to execute active attacks.

TR (Trust Relation) For a given epistemic situation where one is aware of risk or vulnerability V and accepts it, and either T or T* is the dominant factor in the trust relation such that $T \neq 1/T^*$, there exists some constant of proportionality C that depends on V so that $TC = 1/T^*$. See Chap. 6.

Trustworthiness A theoretical definition of trustworthiness can be gleaned from the computational trust literature. A level of trustworthiness is a prediction about the ability of one entity to fulfill the expectations of another entity in a trust relation. Thus, trust can be interpreted as a subjective belief of one entity regarding the intentions or future actions of another entity based on its own past interactions with that entity and those of its neighbors. A more practical definition of trustworthiness can be gleaned from an examination of the application of the concept in electronic commerce. Here, trustworthiness can be defined as the perception of confidence in the electronic marketer's reliability and integrity which is consistent with a virtue-based trust model.

UT (Unified Theory of Trust) Responsible belief is epistemically rational belief and full-fledged trust that is good enough given a context, where the context determines what is epistemically rational and acceptable to believe for achieving some socially valuable end or avoiding some socially undesirable end. See Chap. 6.

Virtue Epistemology (VE) A diverse collection of theories of knowledge which claim that epistemology is a normative discipline, and that intellectual agents and communities are the primary source of epistemic value and the primary focus of epistemic evaluation. VE research programs include novel attempts to resolve longstanding disputes, solve perennial problems, and expand the horizons of epistemology.

Vulnerability A weakness in an asset that can be exploited by a threat to cause harm to a system or device.

Worldwide Interoperability for Microwave Access (WiMAX) A wireless communications standard developed by the WiMAX Forum that enables the delivery of last mile wireless broadband access as an alternative to cable and digital subscriber line (DSL) service.

Wireless Sensor Network (WSN) A network composed of spatially distributed autonomous sensors that monitor physical or environmental conditions such as temperature, sound and pressure, and cooperatively pass their data through the network to a main location. Originally developed for military applications such as battlefield surveillance, WSNs are now used in many industrial and consumer applications. Size and cost constraints on sensor nodes result in corresponding constraints on resources such as energy, memory, computational speed, and bandwidth.

Wireless Local Area Network (WLAN) A network that links two or more devices using a wireless distribution method, typically spread-spectrum or OFDM radio, and usually provides a connection to the Internet through an access point. This gives users the ability to move around within a local coverage area while remaining connected to the network. Most wireless LANs are marketed under the Wi-Fi brand name, and have become popular in the home due to ease of installation and in commercial complexes offering wireless access to their customers.

Wireless Wide Area Network (WWAN) A network that uses wireless technology to span a large geographical distance. A wireless WAN differs from a wireless

LAN by using mobile telecommunications cellular network technologies such as LTE, WiMAX, and GSM to transfer data. A wireless WAN can also use local multipoint distribution service (LMDS) or Wi-Fi to provide Internet access. Since radio communication systems do not provide a physically secure connection path, wireless WANs typically incorporate encryption and authentication methods to make them more secure.

Wormhole Attack An attack that records packets from one location in a network and retransmits them in another location in the network to disrupt the overall functionality of the network.